D1108350

FIREBALLS, SKYQUAKES & HUMS

FIREBALLS, SKYQUAKES & HUMS

Probing the Mysteries of Light and Sound

ANTONY MILNE

ROBERT HALE · LONDON

© Antony Milne 2011
First published in Great Britain 2011

ISBN 978-0-7090-9278-0

Robert Hale Limited
Clerkenwell House
Clerkenwell Green
London EC1R 0HT

www.halebooks.com

The right of Antony Milne
to be identified as author of this work has been
asserted by him in accordance with the
Copyright, Designs and Patents Act 1988

A catalogue record for this book is available from the British Library

2 4 6 8 10 9 7 5 3 1

Typeset in 10/12.7pt Sabon
Printed in Great Britain by the MPG Books Group,
Bodmin and King's Lynn

Contents

Acknowledgements

I am grateful to all at Robert Hale Ltd for their help and advice.

I should also like to thank Gill Jackson, who suggested some improvements to the text, and Hilary Luckcock, the picture researcher.

Introduction

A 'fireball' is actually a vague thing, no more than a glowing sphere, often seen careening across the sky and disappearing behind hills. A fireball is hence a colloquialism, a casual analogy, and not academically recognized as a term that specifically describes meteorite or cometary fragments. It is interchangeable with a 'Ball of Light' (a 'BOL'), and competes with that other vague reference to 'Lights in the Sky' ('LITS'). The same applies to the coined term 'skyquake'. This is an unidentified celestial explosion, possibly a space missile, accompanied by a flash of light. In this case the astronomical term 'bolide' comes nearest to defining the phenomenon.

Scientists have to depend on a classical paradigm to help them explain how and why objects fall to Earth from space, what they are made of, and how much damage they can cause. Astronomers work to a rough tariff that equates the size and density of the space missile with the destructive impact it could cause, measured in megatonnages.

Unfortunately, LITS, in particular, especially because they are a manifestation of the unknown and the undefined, have become fragmented into many different categories. The origins of any skyquake, and the dimensions and make-up of the likely object that supposedly causes it, remain problematic. Fireballs and BOLs raise questions that most scientists do not want raised. The suggestion that fireballs are UFOs often irritates and unsettles them because UFOs are regarded as being in the 'anomalist' category (i.e. they are an 'anomaly'), and ought to be excluded from the scientific frame of reference.

Even so, space itself is too often fraught with all kinds of surreal manifestations. For one thing, the BOLs, if they are not solid missiles, could simply be antimatter explosions in our skies. Alternatively, giant comets could become natural nuclear weapons because the organic nature of their innards, when subject to enormous pressures and temperatures, could make them 'go critical'.

There is another significant factor. In postwar years many flashes and explosions could have been weapons tests, both nuclear and conventional, done both at stratospheric and at ground level. But they too are a challenge to physicists, astronomers and defence analysts. Categories are accepted or rejected, embraced or dismissed, as the explanatory paradigm veers between the respectable and the 'anomalist'.

All astronomers labour under the legacy of something that happened in Siberia in 1908. This was the most famous skyquake event in history. It was a revolutionary turning point that threatened to undermine Newtonian physics. In the absence of an obvious crater and the difficulty in determining how a space missile could cause a massive explosion that was heard for hundreds of miles around, we are simply left with a typical BOL mystery.

Then there were the frequent accounts of rocks crashing through people's ceilings and small stones falling on to their roofs from clear blue skies. This category of falling objects, especially when they included frogs, buttons and beads (complete with tiny holes ready for threading), had ineluctably to be excluded from space and cosmic debris because it was regarded as 'Fortean' (i.e: inexplicable) phenomena. Large chunks of ice were often said to be dislodged from the frozen wings of aeroplanes. But for many years airline companies and a few courageous scientists have rejected this explanation, though it is still insisted upon by newspaper editors who often conceal the fact that the falls of ice land hundreds of miles away from flight paths.

Astronomers have had to cope with yet other phenomena that cross over the Fortean line such as the 'fiery tentacle' nature of many fireballs that landed from outer space, which is also characteristic of UFO sightings. Manifestations of extraneous matter have been reported, and gelatinous substances that tended to evaporate to the touch. Spider-web-like substances that drifted to the surface also made it difficult for scientists to work out whether they were confronted with the normal or the paranormal. The trend, inevitably, was to plump for the former. Similarly, 'chemtrails' were frequently attributed to sinister military aircraft experiments, but equally they could have been conventional aircraft contrails that had taken on a distinctive hue. Indeed, much illuminated phenomena more often than not were amenable to secular explanations. For example, the UFOs allegedly appearing over power stations that had suddenly ceased functioning, plunging whole communities into darkness, were attributed to unusual solar flares that were known to have occurred at the time. Unfortunately, as with surreal cosmic happenings, so with light phenomena. Uncomfortable issues concerning fundamental physics at the level of the photon were pressing: why were some fireballs orange, and others green, blue or white? Monitoring and detection institutions sprang up that tried to separate and list the variously described sightings. But these agencies of enquiry inevitably resulted in considerable overlap, misunderstandings and duplication, even controversy. Some were to be found in academe – such as meteorological and astronomy departments – and some were not. There were official enquiries. Phenomena slotted into some categories but not others, and disturbingly crossed over later in yet others, largely because witness accounts of fireball events so often varied, and video footage could be differently interpreted. For example, an astrophysicist in December 2010 said that two green fireballs seen in Australia were meteors, but were somehow created by cometary debris, and (or but) could also have been the sighting of 'ball lightning', which in turn could have been caused by the original meteor. Here we have ricocheting comets, meteors and ball lightning squeezed into one explanation.

Meteorologists thought they had the best explanations. Indeed it is the balls of light and clouds that are part of the wider array of light patterns that can easily subsume other sky apparitions. Some take on the appearance of shimmering mists or eerie fogs. Noctilucent clouds and sundogs can look like UFOs to the extent that they succeed in blurring the distinction between the natural and the unnatural. These translucent, semi-solid objects, with undefined edges, often sitting motionless in the sky, and virtually emulating clouds, are not uncommon, as can be seen from internet video clips.

Many sightings can also be attributed to our heavily polluted atmosphere. The misty cone of light that sometimes can be seen as the sun rises creates an ethereal glow. Large coloured rings can appear around the moon when it reaches its perigee, and weird light scattering can also be due to volcanic eruptions.

Science confirms that sonoluminescence does occur, as when mysterious flashes of light bubble up from liquids in very low-pressure situations. Dry substances such as sugar give off light by having their electrons stripped away with the snapping of bonds between atoms. Similarly, electrical storms fall clearly into the conventional science category. By themselves, and when they produce a 'ball lightning' aftermath, or when they appear as the strange flickering auroras that can be seen by pilots miles above the stratosphere, are all deemed to be real and explicable.

Unfortunately, lower down, there remains a confusion between 'ball lightning' and other types of global fiery apparition that are sometimes connected with lightning, such as *St Elmo's fire*. BOLs at ground level are often tiny, sometimes football size and pulsating. They can chase cars or appear in front of aircraft or float down aircraft aisles, or crash through bedroom windows. But if they are 'plasma balls' bound together by some form of cohesive energy field it is difficult to understand how they can retain their round shape and keep to surface level, let alone move about seemingly purposely and collide with anything.

It is when they are considered to emerge from the silicon-rich crust of the Earth that again science swings back to the mainstream. 'Earthlights' have long been a respectable interpretation of BOLs and UFOs. The theory dictates that the inner core of the Earth is shaken up by seismic activity which, in a sense, supercharges it so that giant electrical sparks are released into the air.

The similar *will-o'-the-wisp* is typical of the 'marsh gas' type of sighting, but remains problematic because it is again largely allegorical. Phosphorus is seldom found in its pure state in nature, and the vital spark that ignites the flame is not accounted for. Still, the will-o'-the-wisp, so similar to St Elmo's, is a safe stand-by alternative for UFOs or other examples UAP ('Unknown Aerial Phenomena'). The conundrum once more was the similarity with the fleeting, bobbing little lights dodging around country lanes that seemed to possess sentient behaviour. Worse was the fact that these were alleged to be 'spook lights', often associated with ghostly apparitions and connected with the accidental deaths or murders of local celebrities.

In the meantime, mirages are the only UAP that survives intense scrutiny and comes out shining with credibility. Bizarre images were reported of upturned ships, castles in the air, and far-off oceans, and even townscapes, floating on the horizon. These are the natural refractions that occur when different air densities turn the atmosphere into a giant magnifying lens. But there were still sceptics. The Hessdalen lights seen in Norway during the 1980s and 1990s, once regarded as likely mirages from distant reflected natural lights, were later sensed to be some kind of spectral vision, especially when a powerful laser beam was directed at them, only for them to react to it in an intelligent fashion. On the other hand the similar *Min Min* lights seen in the Australian outback, where BOLs seemed to lunge at motorists, were typical mirages of distant car headlights some tens of miles away.

Scientists tried to explain 'spook lights' at sea, which were attributed to 'bioluminescence' from plant or microbe life under the waves. Yet it was a rather desperate explanation. There was more limbo-land wavering between apocrypha and legend, and scientific orthodoxy. The phantom fogs and land masses that occasionally loomed up to threaten shipping were also at first attributed to dense maritime fog banks, and the large and fast-moving 'phantom subs' were real ones that were possibly secret naval prototypes.

The difficulty, of course, in interpreting light phenomena might be simply due to human failings. The limitations of the eye are well known, as the world 'out there' is an ever-changing pattern of electrical impulses perceived inside the human head. Images of unknown people appear in digital pictures when they were not seen by the person who has taken the picture. The 'orb' mystery has still not been explained. Several books and articles show pictures of such orbs, and argue for their existence. Images on the internet reveal that these small circular discs of patterned light, often mottled with green and pink tints, cannot be explained by technical glitches or internal light reflections in the camera itself.

And as digital cameras and camcorders are in the realm of the objective in the sense that they are real, solid items of technology, so are some strange beams of light in the sky also the result of known technology. Some fireballs are the result of 'sky-zapping' exercises done by electrical experiments and defence scientists following in the wake of early twentieth-century tests done by the electrical genius Nikola Tesla. Experimental 'beam weapons' using laser light, radio waves and electrical energy can occasionally shoot out rays that go into the visible spectrum. The HAARP project in Alaska has long been the focus of critics who say beam weapons are being tested in secret.

The most common secular explanation for fireballs – almost the official position – has been the sighting of space debris as it careens towards Earth, causing sonic booms and sky explosions on the way down, and often setting fire to forests. The broken-up pieces of defunct satellites that become molten blobs on their parabolic trajectory to the surface can indeed look like UFOs or cometary debris. Debris can become spaced out, trailing its glowing fragments along a route across the sky that can delude witnesses into thinking that they are looking at the portholes of a giant alien spacecraft. Even satellites still in low orbit can become like mirrors, with their flat antenna panels glinting and flaring in the sunlight. And both natural debris and space junk can sometimes, because of optical illusions, appear to be rising.

As the issue of space junk becomes more of a security and environmental issue, research agencies and committees have started to list and annotate it. But these compilations have to compete with all the others. If there was a danger of over- or under-estimating the nature and quantity of space debris, the growing lists and reports of UFOs written up by defence and intelligence agencies themselves tended to diminish the overall value of debris lists.

As time passed, the mystery of background hums has risen up the agenda. Too often the irritating throbbing noises cannot be traced to their source, but seem to be mainly an urban industrial problem. They encourage sufferers to indulge in similar types of speculation relating to UAPs. Our ears are not capable of accurately assessing and locating the source of sound energies. If unidentified fireballs have been seen hovering over generating stations and have caused blackouts, then those same power stations are often responsible for the mystery hums.

Similarly, if the military have been blamed for skyquakes and beam weapons testing, then the muffled booms often heard in the distance are an aural manifestation of these events. Sounds can be amplified over long distances by air currents in the same way that mirages can be seen as distortions of distant land masses.

Unfortunately, the geophysical explanation has serious limitations, because earth-cracking noises arising from massive seismic stresses would have sharper, high-pitched tones.

The fact that many legends say the drawn-out booms are heard near coastal regions and must hence come from the oceans, or perhaps from under the oceans, is not helpful. The explanation gets pushed further into the distance, literally and metaphorically.

Visions of ancient phantom armies tramping across the countryside are clearly evidence of the paranormal, and fall even beyond the frame of reference that could be applied to interpreting spectal lights. But an open-minded researcher has no alternative but to report these visions as objective phenomena, or at least as *perceived* phenomena, simply out of respect for other people and for the historical record. Witnesses have described, sometimes in detail, the costumes that ancient soldiers have worn that only a specialist historian would recognize. Sometimes the inhabitants of entire villages have been called out to witness these phantom armies as they head towards nearby woods or are seen scaling the hills.

It would be reckless for scientists or psychologists to ignore these kinds of tales. There is now a sizeable literature on psychic and poltergeist activity; people have experienced unaccountable movements of objects that seem to defy the laws of physics and they have heard sounds that defy the laws of acoustics. Many people also seem to have 'extrasensory perception' or have had premonitions, and this has been confirmed in repeated clinical and psychology tests.

The phantom wartime planes seen over British airfields, clearly in difficulty and about to crash, are similar to the apparitions that psychics understand are connected with violent death and tragedy. These stories are fraught with peculiarities, and hint again at the hidden parallel world of alternative futures and constantly recurring pasts. Witnesses see, often in silence, giant B20 bombers just yards above their heads, and workers at aircraft hangars report ghostly sounds of wartime planes being revved up, and hear the phantom voices of dead pilots.

Ghostly voices often echo over the airwaves, something that Thomas Edison and Guglielmo Marconi, both pioneers in radio telephony, have experienced. In the 1960s German psychologist

Hans Bender found faint imprints of human voices left on tape recorders running in silent soundproofed rooms.

UFOs, of course, represent the most obvious clash between straight science and the paranormal, in that they embrace the entire gamut of unidentified aerial phenomena. When they began to be seen in vast numbers in the twentieth century they succeeded in redefining the skyquake condundrum, but blurred the distinction between these and other hidden cosmic forces.

In my own researches into satellite and bolide debris I was only too aware of this awkward overlap, having to report on objects crashing to Earth but at the same time having to cope with unhelpful alternative explanations from sundry astronomers, local people and police. In the meantime the large official literature on UFOs has inevitably to be taken into account. There seems to be not a single defence department anywhere in the world that does not have huge files on the subject. The most alarming aspect, as can be seen on internet clips, is the way that UFOs look like torpedoes or missiles, and fly alongside both military and civilian aircraft.

Many theories about UFOs seem wide of the mark – largely because even ufologists are reluctant to believe that these fleeting, erratic objects, changing their shapes and colours, are earthbound paranormal entities with an insect-like instinctive awareness of their surroundings. They are more like poltergeists than *aliens* – that is to say, creatures with respiratory systems that are travelling across space in vehicles that are alternatively cigar-shaped or flying saucer-shaped, either of which could turn into a BOL in an instant!

Nevertheless, recent advances in physics at the atomic level can help explain how a sentient light source – a living light source – can utilize millions of kilovolts of energy that can be gleaned from the space environment itself. UFOs might even be able to use the giga-electron volt energy housed within the mass of the proton to deploy anti-gravity technology, or to deploy 'invisibility cloaks'. All this is theoretically possible, and in some cases achievable by human scientists. There are some extraordinary advances in LED technology that show how light signatures can be manipulated, UFO-style. Physicists know that gaseous substances can take on the appearance of solidity in certain circumstances. Even electrons are sometimes known to seem like liquids and at other times like solids. Perhaps UFOs know how to gain mass on occasion, and at other times to lose mass and become invisible.

We must simply admit that we are limited. It will be a long time before humans gain a real understanding of UFOs because we have no real knowledge of the nature of reality. UFOs remain a mystery because fireballs remain a mystery. The Jack-o'-Lantern, ball lightning and the spectral light must be regarded as part of the Whole, where every part of the universe contains the whole order. Some ufologists suggest that UFOs might themselves be holograms, and be able to switch on an invisible laser beam that can change its appearance at will.

If the phantom armies are merely glitches in the fabric of the universal hologram, then so are the fireballs and the skyquakes.

What is the abode of Light? And where does Darkness reside?

The Book of Job 38:19

A philosopher once said, 'It is necessary for the very existence of science that the same conditions always produce the same results.' Well, they don't!

Richard P. Feynman

1 The 'Fireball' Riddle

Throughout May and June 1993, in remote scrubland in south-western Australia, mysterious loud rumbles and explosions could be heard at ground level for miles around. If this was a missile from space, then its nature, its fiery hue, its speed and its trajectory, seemed to vary according to who was describing the phenomena. Even stranger was the fact that the object causing the first explosion at surface level seemed to be on a ballistic-type Earth-hugging trajectory, like an exceptionally fast cruise missile. Many could observe its flight path over a distance of more than 180 miles. It appeared to arc down towards the ground before disappearing behind low hills. Three further related events occurred in quick succession, with minor quake damage reported as far as 1,100 miles away.

Harry Mason is an Australian geologist with extensive field experience in geological prospecting and mapping with computers. He was surprised to read press stories in the early southern winter of 1993 of a regularly returning 'meteor fireball' that flew from the south between the two Western Australian towns of Leonora and Laverton. Leonora is situated 146 miles north from Kalgoorlie, which is about 515 miles east of Perth. Both Leonora and Laverton are now attractive small towns in the 'Eastern Goldfields' region, and retain an early twentieth-century housing style.

Most of the inhabitants of these communities believed that a large cosmic missile had set off an explosion, or that a chemical or fuel dump inside a gold-mine had been violently detonated, and that other missiles followed to create more explosions. Mason began documenting witness reports of these peculiarly disturbing events. He located an engineer who said that the blast wave was 'conclusive, and much bigger than a normal open-pit mine blast'. A series of several silent fireballs, travelling separately at regularly spaced intervals, were seen a year later in October 1994 in the Pilbara region of Western Australia. These were mostly large and red-orange in colour, and were travelling at aircraft speed at a very low altitude in full view of about 4,000 people. Some said they looked like 'balls of plasma', and others said they had a 'central black hole'. There had been many other similar Australian reports during the rest of the 1990s – as many as 1,000.

These dramatic fireballs and the earth-shattering explosions that followed them, in virtually every case, mystified Mason and a good many other scientists. In fact the events in Australia raised a great many scientific questions concerning Earth and space. Gregory van der Vink, of the US Incorporated Research Institutions for Seismology (IRIS), hinted originally that meteorites might be to blame for all these sightings in Australia. But one object changed its colour to blue-white and then took off at high speed, proving that it could not have been a meteorite. In other cases there were reports of astronomically (indeed aeronautically) impossible sudden changes of course.[1] Further, no one referred to sonic booms in the lower atmosphere (as opposed to the surface), which was puzzling if they were solid missiles that had indeed entered the atmosphere at over 1,000 mph.

The peculiarity of the events gave rise to speculation about UFOs, an aspect of the fireball mystery that will be given considerable sympathetic treatment in these pages. For one thing, puzzled investigators could find no trace of a crater in Western Australia after contacting Sandia which is a National Laboratories major science, aeronautics and engineering complex in New Mexico. Seismic readings were, perhaps, picked up by Ed Paul, a geophysicist at the Australian Geological Survey Organisation (AGSO) near Perth. He suggested the first main 1993 event generated the energy equivalent of 2 kT of TNT, and said that the airburst-explosion over Western Australia, or any similar shock wave, would indeed be typical of a meteor impacting.[2] Columbia University seismologists said that after-shocks from the event hinted at a quake, and not a bomb or missile. The earthquake theory was also advanced by seismologists Christel Hennet of IRIS and Danny Harvey of the University of Colorado in Boulder, who said that seismic readings hinted that so-called S-type sheerwaves had occurred, a sure sign of an earthquake.[3] Harry Mason, the investigating geologist, remained unconvinced, since the events failed to fit in with any earthquake phenomena that he had known about. He also pointed out that for any meteorite activity to occur repeatedly within a six-hour period would be unlikely.

Bolides as nuclear missiles

There have been literally thousands of astounding and mysterious explosive fireball incidents in recent decades, and the most widely commented-upon incident, analysed in countless books and articles, is the Tunguska event. It has become one of the world's most intriguing unsolved natural mysteries. This awesome disaster occurred in the subarctic coniferous region of Siberia in 1908.

A 'bolide' does describe the Tunguska event accurately. 'Bolide', which comes from the Greek word 'bolis', can mean a missile or a flash that explodes. The International Astronomical Union generally considers bolides to be synonymous with very bright fireballs. Geologists often refer to bolides as missiles from space, leaving large craters, the nature of which – meteorite, asteroid or comet – is unknown. Whatever caused the explosion in that remote part of the world, heard for miles around, no evidence of its original form has ever been found in the region. It came, it destroyed and it vanished, leaving no evidence in the form of fragments, nor any sign of a crater, exactly as in the Leonora/Laverton case. Oddly, though, the Tunguska disaster caused high winds that could be felt 300 miles away. Much of the *taiga* had been unaccountably scorched, and the tops of the trees had been clipped off. The blast could be heard in Kansk, 500 miles distant. Massive tsunami waves crashed into the banks of the Angara River, in the south-east of Siberia. Eerie glows lit up the night skies over Europe for weeks afterwards.[4]

There was one vital clue. The explosion had injected massive amounts of water vapour into the dry upper atmosphere. This hints that the impact was caused by a wet, icy comet, rather than a meteorite. According to Michael Kelley of Cornell University in Ithaca in New York, the vapour was carried by high-speed winds into the thermosphere – about 190 miles up – to be seen in the western hemisphere within twenty-four hours, hundreds if not thousands of miles away.

If asteroids can behave like ballistic missiles with conventional warheads, then comets in particular – because of the more organic nature of their innards and because of their (apparently increasing) size – can instead become virtual nuclear weapons. In eastern Siberia, north of Lake Baikal on 25 September 2002, a Tunguska-type object destroyed 40 square miles of forest and caused earth tremors. The Russian media reported that villages 60 miles away had witnessed a 'gigantic' fireball screeching down from the sky, causing windows to rattle. It was thought perhaps that 'fragments' of the meteorite had exploded with shrapnel 18 miles above the Earth, with the force of at least 200 tons of TNT.[5] And a bolide that landed in Tahiti in the year 2000, according to science journalists, released as much energy as a small nuclear bomb, about 3 kT of TNT.[6] (The nuclear weapon that destroyed Nagasaki in 1945 was the equivalent of 20 kT).

But the nuclear analogy only works if one is referring to a *fusion* bomb. Whereas fission involves the splitting of the nuclei of heavy isotopes like plutonium, fusion (as in the hydrogen bomb) involves the joining together of light atoms. Nuclear fusion takes place when the hydrogen nuclei are subject to very high temperatures and pressures – similar to what goes on in a catastrophically descending bolide. A comet would, once its outer nucleus is worn away, have its tightly packed volatile gases – consisting largely of hydrogen, deuterium, ammonia and helium, trapped inside a shell of some sort – heat up catastrophically. Its gases, normally held back by the brute force of the front-ended shockwave, would try to escape through the molecular shell. But the cross-section of any fast-moving body from space tends to increase. If it has an outer stony shell it enters the Earth's atmosphere at something like 10,000 mph. On hitting the air it would be instantly decelerated by enormous kinetic friction. The air would become compressed and heat up to thousands of degrees. If the pressure is raised to 125 atmospheres (1,800 pounds per square inch), the temperature rises dramatically to that required to detonate heavy hydrogen or deuterium (about 9,032 degrees F). Soon the cometary gases would blast violently through this shockwave. In other words, it 'goes critical', and a kind of thermonuclear reaction could occur a few miles above the ground, utterly destroying the object. It would cause blast damage and terrific explosions without actually thumping into the Earth itself and causing a crater. This would explain the suspicious absence of damage to the Earth's surface in so many cases, and which so often ends with a fruitless search for the missile.

To support their idea of a Tunguska fusion bomb, evidence of radioactivity in the region was produced by Soviet scientists in 1969. They discovered that forty-five-year-old trees that were saplings at the time of the 1908 event had attained much greater heights and girth than would normally be expected. Other scientists at the Russian Geographical Institute of the Academy of Sciences found that many of the trees in the region showed tree-ring evidence of radioactivity, even evidence of artificial radioactive isotopes.[7]

A somewhat different scientific angle was recently put on the nuclear issue. Comet Holmes, nearly 2 miles across, itself exploded with nuclear-bomb ferocity somewhere in space in 1997, and became a million times brighter, according to the *New Scientist* magazine. But this event, we are told, could have been caused by a freak (and unlikely) collision with an asteroid. William Reach of the California Institute of Technology in Pasadena thinks the culprit may have been an exotic and unstable form of water ice at the comet's heart. This is because in sub-zero temperatures in outer space the crystal shapes of water molecules get stuck together much more haphazardly. Only with a slight rise in temperatures, which would still leave the missile at -207 degrees F, does it revert to its crystalline form, squeezing out gases trapped there. This causes massive instability, resulting in a high-energy explosion.[8]

Mini black holes?

Could the strange light-forms seen in recent years that have resulted in destructive blast damage to the surface of the Earth have been due to a collision with a 'mini black hole'? A black hole is in effect the celestial phantom of a star originally many times bigger than our own sun. It gets into a bizarrely shrunken state of affairs when its matter collapses in on itself and contracts beyond the point of no return. This is because in the process of shrinking to merely a few miles in diameter, its phenomenal power of gravity remains within it to overwhelm every other force in the cosmic neighbourhood. Nothing would then stop the collapse to a pinpoint. It would devour its own body and everything around it. Even the light from its own tiny white-hot nuclei would be trapped and unable to escape this irresistible force.

Soon the concept of a *mini* black hole emerged – potentially lethal, but not big enough to destroy a planet. Some believe that mini black holes were commonplace some 14 billion years ago when the universe was 'created'. This arises from the suspicion that unlike the massive black holes that form when a star explodes, the mini version would form wherever matter accrues in dense clumps.

Until enough is known about black hole monstrosities from outer space it is difficult to rule out the possibility that pin-prick black holes are responsible for the smaller fireball sightings on Earth, resulting in loud explosions and illuminated phenomena.

Mini black holes could be raining down through the Earth's atmosphere right now. They could bounce in and out of a speeding asteroid before coming to rest at its centre, and just quietly devour it atom by atom, thus resulting in their unusual behaviour and trajectories. Shrunken objects attacked by black holes would start to glow and melt, and would become immensely heavier. In 1931 a 'meteorite' event that occurred in South America, at Rio Curacao, on the Peru–Brazil border, didn't seem to have the usual 'footnote' of a meteorite impact because a forest caught fire and apparently blazed for several months. A similar event happened in British Guiana in 1935, where trees were flattened for up to 20 miles around.[9]

Seismologists, such as Paul Rickard of Columbia University in New York City, are beginning to take this seriously as an alternative explanation for seismic activity. He and others believe that a mini black hole would cause a definite crater-like impact, even deep seismic disturbances. To produce the kind of energy released at Tunguska, a black hole smaller than a speck of dust is all that is required. 'It would be ten micrometres across, weigh about a ton and be going 400 km [250 miles] an hour', explained scientist Eugene Herrin.[10] On its way through the atmosphere the mini black hole would draw in constituents of air from a wide area round itself. Its temperature would rise to between 30,000 and 300,000 degrees F, emerging from ultraviolet radiation that would produce a plasma column that would appear as deep blue. This could explain the mysterious illuminated observations sometimes reported from mountaintops.

Much of this, of course, is erudite speculation, and presents us with as many metaphysical problems as scientific ones. Atoms under attack from mini black holes would explode in a burst of particles within one-billionth of a second – the kind of atomic instability that scientists normally associate with plasma or cosmic rays. On the other hand, it is not only the black hole that has 'massive gravity' – so does the nucleus of an ordinary earthbound atom. Electron capture by atomic nuclei is limited if the nucleus is not in the right quantum state to convert a proton into a neutron so that the gravitational field of the mini black hole would be much smaller than the binding energy of electrons in atoms. Mini black holes, in other words, would have to compete with the atomic nucleus – and its strong nuclear force – with its own gravity trapping atoms further out.

There are other theoretical problems. Scientists know that black holes would ultimately leak energy into space. A black hole weighing a million tons – a tiny pinprick of dense matter – should evaporate within 30,000 years. So if big black holes evaporate via radiation, then atomic-sized black holes are even more likely to do so.[11] In any event they would probably be deflected also from the Earth's magnetic field, and thus have even more difficulty in penetrating the Earth's air space.

Antimatter and fireballs

An alternative theory is that the Australian and Tunguskan fireballs could have been the result of a collision not with a black hole, but with *antimatter*. Explosive antimatter energy is now known to be greater than any other source known to science. Six cubic inches of antimatter crashing to the Earth's surface would flatten huge areas, and could take the lives of millions. Atoms of matter and anti-matter are constructed in precisely the same way, and have the same atomic weight, but the electrical charges are reversed.

Electrons would be positrons (positive electrons) and vice versa. This would apply also to neutrons, protons and a host of others. If the two atoms should meet in space they would vanish in a tiny but intense flash of gamma radiation. In serious bolide-impact events, it is possible that antimatter did cause explosions soon after entering the Earth's atmosphere.

We could, in passing, ask why, if matter and antimatter were created at the beginning, did they not then annihilate each other? And why, for that matter, don't they annihilate the entire universe right now? Perhaps there were subtle differences in the physics of matter and antimatter at the beginning that are not yet understood. One theory suggests that gravity affects antimatter differently, and this explains why the antimatter flew apart at the beginning. Perhaps the antimatter is in a separate region of the universe. This could inform scientists who are still searching for a 'gravity parti-cle' that could explain why gravity seems to work in some areas of space, but not in others, and why it seems occasionally to push rather than pull in some other weird cases seen only in laboratory experiments.

One particular idea can perhaps explain how the imbalance springs from the characteristics of the weak nuclear force, which govern certain nuclear processes, including radioactive beta decay. In the 1960s physicists found that the weak force is not quite symmetrical in regard to matter and antimatter, making the laws of physics seem lopsided, rather like the imbalance between matter and antimatter itself. In 1998 the CERN particle physics researchers in Geneva showed that one particle, the kaon, turned into its antiparticle slightly more often, creating a tiny imbalance. Another idea centres around a particle called the *majoron*, which is thought to have created neutrinos and antineutronos – tiny mass-less particles even smaller than the electron – but not in equal amounts.[12]

A word of caution again: these little antimatter warheads could be lethal if emanating from our own solar system. But they could not possibly retain their destructive power for long if they emanated from far beyond our own galaxy. Even travelling at the speed of light the antimatter particles would have taken several million years to get here. On reaching our solar system, they would have been whittled down by countless minor explosive impacts with interstellar particles of ordinary particulate and gaseous matter. They would, finally, very likely be annihilated when trying to reach the Earth's surface after hitting the atmosphere, which would contain gases that would be millions of times denser than anything encountered in outer space.

Still, size matters. Even a small piece of antimatter no bigger than a pea, should it survive its galactic journey to Earth, is all that is needed to cause the Australian fireball blasts and piercing lights. This is because the antimatter impact would, by clashing with and destroying positive matter, produce pure energy. This could be evidenced by the brilliant blue streak seen in Siberian skies in 1908, and is precisely the visual effect a chunk of anti-matter would produce as the particles reacted with the dense positive molecules in the atmosphere. They would at least cause a mid-air blast, especially if they produce gamma rays in the process.

And these gamma rays themselves could arise from other aspects of space. These are fast-moving and highly penetrative rays at the shortest possible wavelength, and probably come from supernova explosions or from the collision of blobs of matter with other matter, or with twisted magnetic field lines that are constantly zapping throughout space. When a gamma ray hits an atom in the upper atmosphere it dislodges a cascade of electrons that generate tiny flashes of light called Cerenkov radiation. One Nasa RHESSI satellite reported fifty gamma ray flares within the atmosphere every day, many with massive electron-volt energies.[13]

An exceptionally powerful burst of gamma rays was said to have struck Earth on 27 August 1998, resulting in strange glowing light effects. 'It was as if night was briefly turned into day in the ionos-phere', said a Stanford University scientist in regard to one gamma ray event.[14] A huge galactic flash from halfway across the galaxy acted on Earth's ionosphere in late December 2004. The eruption in space was noted by the US National Science Foundation's Very Large Array of radio telescopes, as well as by satellites. This was caused by a *magnetar*, a neutron star with an enhanced magnetic field, and was 100 times more powerful than any other similar supernova or gamma ray eruption, according to David Palmer of the Los Alamos National Lab at New Mexico.[15] 'Had this happened within ten light years of us, it would have severely damaged our atmosphere and possibly have triggered a mass extinction,' said Bryan Gaensler of the Harvard-Smithsonian Center for Astrophysics.

'Skyquakes' around the world

Some of the fireball features, as described by witnesses, are truly awe-inspiring. Many are clearly overly dramatized and embellished. Some witnesses speak of 'fast-moving stars' and 'dazzling daylight suns'. It is often difficult to take these reports seriously, but scientists often confirm they do happen on this scale. Take the object that exploded over northern Sudan on 7 October 2008 and that lit up the sky 'like a full moon'. This was caused by a 10-foot asteroid, which disintegrated in the atmosphere; even an airliner some 870 miles away saw the flash. It was also spotted by Nasa's observatory in Arizona that was conducting an asteroid-watching scan,[16] but the reason why its disintegration should cause such a bright illumination was avoided in the scientific literature.

During the winter of 1977–78 loud 'skyquake' explosions emanating from the upper atmosphere shook the eastern USA from Connecticut to South Carolina. But there is a controversy about whether these prolonged booms in the air are actually meteorites impacting in the atmosphere, sonic booms from planes, or something else entirely. People living in New York State, over many years, have heard loud booming noises, and if they weren't attributed to meteorites they were regarded as the result of seismicity, such as the periodic cracking and grinding of rocks, as it was commonly expressed.

Other US states had experienced similar booms and rumbles that rattled windows and sometimes set off scientific instruments measuring air pressures. This happened in Charleston, South Carolina, and continued in a kind of chain reaction to the Canadian province of Nova Scotia. According to William Corliss, who wrote a detailed account of strange atmospheric noises in 1983, 'On occasion, flashes of light have accompanied the booms. Explanations for the strange sounds have ranged from supersonic planes to methane gas bubbles that have been ignited by static electricity, but no conclusive evidence has yet emerged ...'[17] (See Chapter 12).

A study by the US Naval Research Laboratory concluded in the late 1970s that a combination of unusual weather and supersonic military aircraft was to blame. It pointed out that similar sounds in Europe were blamed on Concorde. But this was contradicted by William Donn, of the Lamont-Doherty Observatory, in Wilmington.[18]

But were unusual weather and sonic booms also to blame for the loud bang that was heard during the night over North Wales on 23 January 1974? Unusual lights were seen for hours, and 'brightly coloured' objects were reported streaking across Welsh skies. Although some of the sounds were attributed to supersonic aircraft, many of them remained unexplained.[19]

Other reports of exploding fireballs

1997   On 16 September, on the campus of the University of North Carolina at Chapel Hill, a boom registered 1.1 on the Richter scale. Again sonic booms were ruled out.[20]

1997   On 17 December a huge aerial blast fractured windowpanes in Rogersville, Missouri, some 13 miles east of Springfield. The Air Force denied any involvement.[21]

1998   Thousands of homes were rattled by two huge booms some 30 miles south-east of Los Angeles in May. It was described variously as either due to earthquakes, or 'explosions', or just 'thuds'.[22]

1998   In August, two loud skyquakes startled 'hundreds' of people on the beaches of Ocean City, Maryland. There were no planes seen in the skies at the time and the booming seemed to come from some miles offshore.[23]

1999   In January, a loud boom was heard at 12.15 a.m., and disturbed residents of Colorado Springs and Denver. Some said it was accompanied by a flash of light in the sky. Again the military denied any involvement.[24]

2000   There was a massive asteroid airburst over the Yukon in northern Canada early in the year. The blast, which took place 9 miles above the Earth, was equivalent to a '4,000-ton bomb going off'.[25]

2000 Just after Christmas, a small meteorite caused sonic booms and streaks of light over south-east Australia, sparking off calls to the police.[26]

2001 On 27 October a gigantic flash and a series of explosions were heard all along the UK's east coast, said to be the meteoroid fragments that followed close on the tail of a comet, but was, strictly speaking, a bolide.[27] According to a British Geological Survey team, a bang was reported, and houses were shaken in the Shetlands.

2003 On 16 June, a loud explosion over Japanese skies left its signature in the dense network of seismic stations that scatter the country. A space rock was said by scientists and the media to be probably travelling at 10 miles per second, and approached the ground at a shallow angle of 15.5 degrees.[28]

2009 On 1 August, a boom was heard throughout Narragansett Bay, Rhode Island.[29]

2009 On 8 October, an asteroid detonated 12 miles above South Sulawesi, on the southern peninsula of Sulawesi Island, in Indonesia. It released, according to the *New Scientist*, 'as much energy as a 50 kT nuclear bomb' but was only about 10 yards in diameter.[30]

These fireball events, of course, can cause fear and panic, not least in the top echelons of society. Some explosions could indicate weapons testing. In the spring of 2010 residents in Cornwall had a power cut shortly after fishermen reported a huge 'blue electric flash' that lit up half the sky. The source was below the horizon, and it was suggested that it was due to frigates and warships testing weapons such as electro-magnetic-pulse (EMPs) devices or mines.[31]

During the Cold War most test explosions were conducted at an altitude that was no higher than 30 miles. Not only was the heat felt by personnel on the ground, but the explosion also became the equivalent of an EMP, and could have been used, and possibly was, as an anti-satellite weapon. US tests caused power cuts in Hawaii, while Soviet tests destroyed phone and power lines in the USSR. In one instance, a power station caught fire and was burnt down.

When Texans heard several loud boom-like noises to the north on the afternoon of 20 January 1994, rumours also began to circulate about weapons testing, and possible disinformation from official agencies about the true nature of the noises was rife. Criticism was levelled at misleading remarks from a supervisor of a section concerned with satellites at the Federal Emergency Management Agency (FEMA).[32]

Dubious rumblings in August 1997 on the remote Russian island of Novaya Zemlya, which was once used as a nuclear weapons testing facility, also caused concern. A 'tremendous bang' was heard over 60 miles distant in the district of Nova Brasilândia, in the Brazilian state of Mato Grosso just a month earlier, and was also suspected of having military connections. In both cases witnesses closest to the events spoke of explosive, 'roaring' sounds.

The space surveillance process – in particular, the United States Nuclear Detonation Detection System – can assess ambiguous events that could be seismic in origin to determine whether they could be nuclear blasts. Important clues come from military satellites, many operated by the US Department of Defense and the Department of Energy. A rogue nuclear test can be distinguished from a meteorite strike by checking the amount and characteristics of the radiation the blast gives off, or doesn't give off. The sound wave of an underground nuclear explosion can travel up to 620 miles from the site of its impact.[33]

As a general rule, scientists can point up the nature of an explosion by combining infrasounds from several nearby stations, and can tell the difference between a point source and a moving source – like a meteorite – and whether the blast might have come from an underground site. Space-based explosions differ greatly from those on the ground: the explosion would appear spherical, rather than mushroom-shaped, although eventually this ball of light would gradually degrade into something approaching the colourful sheens of the aurora borealis (see Chapter 5).

The Sandia National Laboratories annex in Livermore, California, uses satellite images to search for surface craters left behind from the earth-crumping effects of underground nuclear tests. Often it is easier to detect a missile impact, because space sensors can more easily track even momentary flashes produced by small asteroids in the upper atmosphere from their visible and infra-red sensors.

The changing colour spectrum

Meteorites can vary from *white*, at their least hottest, through the colours of the spectrum, ending up with a deep *red* or *pink* as they decelerate. Green, red, blue and white seem to be the most common colours for asteroids and meteorites, depending on their speed of entry.

What is puzzling is the 'pecking order' of some aerial lights, with the bigger orange ones – presumably the first to approach the Earth's surface – chasing the smaller bluish ones after having done a smart U-turn. This seems to be the reverse of what should happen. The colour spectrum accorded to bolides raises many questions about the trajectory and composition of the observed space objects, and even the way the human eye perceives colour (see Chapter 8). Various lights ranging from *bluish-white*, '*whitish*' to *yellow-orange*, with bigger ones chasing small ones, were seen by eighty-six eyewitnesses at different localities and times in the east Devon area in the UK in August 1997.[34] These were logged by a UFO group, who said the sightings had nothing to do with air traffic from Exeter airport or a local hot air balloon group, nor with military exercises on Woodbury Common, as had been suggested.

On 9 October 1997 a meteor loudly impacted the West. In one meteorite case that occurred in Bulgaria, its extraordinary brightness was alleged to have exceeded that of the full moon, although the breakaway chunk only caused momentary *blue-white* lights. In another case of crashing meteorite debris, witnesses spoke of 'artillery fire and boulders falling from the sky', with each fragment detonating individually.[35]

Take the unnerving incident that occurred on 23 January 1974. Scores of people from the Berwyn Mountain area of North Wales and Cheshire phoned Colwyn Bay police to report a mysterious object which had a 'long, fiery [presumably *red-orange*] tail'. It was accompanied by a loud explosion, and landed on a remote peak. The Anglesey Coastguard did report a fireball, but it was, he said, *green* in colour. Coastguards and officials in Lancashire and Cumberland also reported green flashes, which were put down to a 'meteor shower'. Confusion abounded when this was suggested. RAF Valley in Anglesey were alerted, as was air traffic control at

Preston, Lancashire. While this event coincided with an earthquake that occurred in the area, a magnitude 3.5, it was still considered to be the explosive aftermath of a meteorite impact, although no debris or crater was found despite searches by astronomers from Leeds, Liverpool, Durham and Keele universities. In 1982 Andrew Pike, an astronomer closely connected with UFO research, wrote a report on the mystery for the Royal Astronomical Society. His findings suggested that RAF night bombing exercises, accompanied by 'photo-flash exercises', could have combined with a possible quake.[36]

The deputy governor general of Kerman Province, in Iran, said on 8 January 2007 that a 'radiant UFO' had possibly crashed in the Barez Mountains. He said the crash was followed by an explosion and a 'thick spiral of smoke'. This was, of course, a classic 'fireball', but there were rumours that it was a US spy drone that had crashed. Another Iranian fireball was reported just two weeks later by Fars News. This was again classified as a UFO in some versions of the report. It occurred in western Iran, which covers the provinces of Khuzestan, Kohgiluyeh and Boyer Ahmad. It was observed for a whole hour, which rules out a bolide. It was described as being 'yellowish' and 'reddish' in the centre.[37] FARS noted that people in the city of Rafsanjan had witnessed a similar incident several days earlier, and 'similar crash incidents have been witnessed frequently during the last year all across Iran, and officials believe that the objects could be spy planes or a hi-tech espionage device'.[38]

Other objects seen in the longer reddish frequencies

1997   The meteorite that hit Shandong province in eastern China in February was said to have turned the sky *red*.[39]

1997   On 3 October and again on 5 October, both in the late evening, the sky over Liverpool was similarly lit up by an *orange* meteorite passing overhead.[40]

1998   Thousands of eyewitnesses saw a 'very bright *white* light' tinged with *orange* fly over the Sierra Madre Occidentale mountains in north-west Mexico in December.[41] Other witnesses reported seeing a number of falling objects plus explosions, and described *white*, *orange*, and *yellow* trails of smoke at this time. But in many cases this colour sequence, according to eyewitness reports, is reversed.

2003   The descending *red* lights seen off the coast of Stonehaven, Aberdeenshire, on the evening of 13 November were thought at first to be distress flares, but a Coastguard spokesman said they – or it – could have been a 'shooting star'. Yet *stationary* red lights were also seen that same morning over Aberdeen airport at 5.30 a.m.[42]

Some scientists believe that any UAPs in the *blue-green* spectrum can be ruled out as bolides, even though this hue regularly crops up in reported fireball events, which implies that they must be some other type of phenomenon. So what were the strange fireballs in the white/green/blue spectrum that were seen across the world in the 1990s? In the San Luis Valley, south-central Colorado, beginning on 30 November 1993, and continuing until the end of December 1994, several six-week periods of local explosions were heard and coloured fireballs and 'orbs' were seen by hundreds.[43] In the middle of this period a NORAD official contacted the Rio Grande sheriff to report a 'significant explosion' at Creek Canyon, Colorado, and complained of glowing green 'giant' shapes streaking overhead. These were seen descending into the San Luis Valley.

In December 1994 a meteor was detected entering US airspace by six US spy satellites and was again described as a *white* 'glowing fireball'. Indeed, officials at Nasa said they had received an unusually high number of 'white light' sightings throughout the whole of 1997.[44] Sceptics doubted the suggestion that the accompanying 'booms' were related to meteorites or perhaps to military activity.[45]

In the UK, the emergency services throughout the South West and Midlands were inundated with calls after a bright, *bluish* fireball arched low across the horizon on the evening of 15 March 1998, when it was already dark. Just three months later police forces in eight counties received calls from hundreds of witnesses who saw bright *blue* lights travelling from the Devon area in a northeasterly direction, again at night, but others who witnessed the final descent stage say they saw only *red* colours. Some witnesses saw a group of smaller lights (presumably the disintegrating fragments), while others saw only one light.

Thousands of people in Spain called the emergency services after seeing 'balls of fire' of an unknown hue in the sky in late 2003, accompanied by earth tremors and sonic booms.[46] Investigations and searches were made by the police of the suspected area in northern Spain, over Madis and around the coast of Valencia in the east. The Spanish National Meteorological Institute said that its radar and satellite images showed nothing abnormal, although firemen were called to extinguish several blazes. Jose Angel, director of the University of Santiago de Compostela's observatory in Spain, said the sighting was probably a meteorite. Possibly the red spectrum may have confused some witnesses when the effects of the sun's rays penetrated into the atmosphere.

What was the brilliant flash of light seen over Greenland early in 1999 during its almost perpetual dark season?[47] Seismometers recorded a ten-second shock, so the object must have landed. Reports from trawlermen, corroborated by videotape evidence, said that a flash lasting two seconds was seen. One witness spoke of 'a very strong light rolling down from the air. It was like a circle burning, and the air around the circle was very light *green*.'[48] From the descriptions of the size of the dust and steam cloud caused by the impacting Greenland object, some five billion tons of ice were estimated to have been vaporized. Astronomers from the Tycho Brahe Planetarium in Copenhagen said the object could well have weighed a few million tons.

In the Western Australian case I outlined at the beginning of this chapter, we saw that the red spectrum was the most common, until the object 'landed', when blue predominated (original reports also spoke of green tints). Does this confirm or rule out a meteorite? A fireball event occurred some days later after the main multiple event, possibly in early June 1993. Several truck drivers reported that, very early in the morning, a 'moon-sized' fireball flew south to north at a low trajectory. Some reports said that it was yellowish orange in colour, with a tiny blue-white tail. Yet others, the majority, said it lit up the dark morning sky with an intense bluish light, the colour of the tail, as it headed vaguely westwards from Laverton to the isolated Banjawarn sheep station. This bluish hue, with the blurred outlines of both the object itself and its atmospheric tail, began to dominate press reports.

Other objects seen in the shorter (blue/greenish) frequencies

1962 On 18 January, a streaking object seemed to cross the entire continental USA, and was seen by 'thousands'. It had a 'brilliant yellow tail flame'. The object was initially explained away as a bolide, but this explanation became rather suspect to many after it apparently crashed near Eureka, Utah, interrupting the town's electricity supply, before – as in the Australian case – taking off again. It then headed over Reno towards Las Vegas, Nevada. Dr Robert Kadesch, a physicist from the University of Utah, insisted it was indeed a meteor, saying that it exploded some 65 miles up in the stratosphere, and presumably did not 'crash' as was alleged, and possibly the other sighting of a flying object had a different explanation.[49]

1969 A brightly coloured *greenish-yellow* object was seen to flash across many counties of England on its way from London to Lancashire and Nottinghamshire one evening in April. Astronomers at Jodrell Bank said it was a meteorite, but astronomers at the Greenwich Observatory said it was merely Russian space junk.[50]

1994 A daylight meteor was seen in May as an elongated fireball and was photographed over the skies of Grand Teton National Park, Wyoming. It reportedly plunged towards the outer atmosphere at over 50,000 mph – and reached 8,672 degrees F, glowing *white* hot. It was accompanied by sonic booms loud enough to rattle windows.[51]

1996 On the Isle of Lewis on 26 October, in north-west Scotland, 'bright' LITS and explosions were heard, and spiralling debris was seen falling into the sea accompanied by smoke trails.[52] RAF Fylingdales said no space debris had entered the atmosphere in the Outer Hebrides region, although others, like the coastguard service, did not rule this out. The British Geological Society recorded sonic effects from a region some 100 miles north of Edinburgh.[53]

2000 The *greenish-blue* colour was also seen after an explosion high in the atmosphere near the US–Canadian border in January.[54] People living in the Alaska and Yukon areas saw the morning sky suddenly lit up by a turquoise streak of light. Some said it was 'as bright as daylight', and produced a shock wave felt 10 miles away. US defence surveillance satellites picked up the flash, which hinted that the energy released was equivalent to about 5 kT of TNT.

2003   Astronomers concluded that a rare *bright-green* meteor flew across Auckland, New Zealand, skies on 12 June. Stardome Observatory spokesman Warren Hurley said the meteor was particularly bright and larger than average. It was visible for ten seconds as it travelled along the east coast of New Zealand.[55]

2007   On 1 January, a bright light was seen from the mountainous Lleyn Peninsula in north-west Wales, and it too was probably a meteor, according to a BBC News website. The light the object gave off was apparently a 'long line, thicker at one end, *bluey-green* and flashing'. Jay Tate from Spaceguard UK said it was probably space debris.[56]

2010   A journalist who spoke about a fireball event that occurred in April in Cornwall said there had been a huge '*blue* flash'. He also said that he had seen a *green* fireball, moving more slowly than a meteorite, which left a faint emerald trail behind it. Then the object swerved 'violently' to the left, before making a faster ascent, and then 'winked out'.[57]

If the colour spectrum of flying objects is puzzling, so is their actual *nature*. The differences between all types of natural exploding missile (a 'bolide') might not be all that great. Further, the size of space objects, and their nature, and the number of them up in space and threatening to land on Earth at any moment, remains yet another cosmic mystery. So, for that matter, do the peculiar 'tentacles' and fiery appendages that some fireballs seem to possess. Even more puzzling are the strange objects that have fallen through people's roofs that may or may not have come from space, as we shall see in the next chapter.

2 Pummelled from the Skies

I n recent years Nasa has succeeded in presenting the astronomical world with a genuine puzzle. The Nasa NEAR Shoemaker spacecraft revealed that large asteroids, like the 20-mile long Eros, are solid rocks with a similar density to the Earth's crust, but others, Nasa pointed out, such as Mathilde, have low, uneven densities, not much denser than water – rather like a giant beanbag.[1]

Indeed, recent scientific discussions of asteroids tend to dramatically reduce or increase their numbers, as space rocks fall first into one category and then out of another. Further, the nuclei of most comets entering the atmosphere have been disputed: some believe that they are just a heap of rubble mixed with icy particles, held loosely together. Whatever they are – and this has surprisingly not dawned on the general public – they can be truly *massive*. In January 1997, thousands of people watched Comet McNaught streak across the skies of Australia, New Zealand, Argentina, Chile, South Africa and Scotland. According to *The Times*, it had a 'bigger head than Mount Everest'.[2] It shone brightly, although it was supposed to be 'millions' of miles away from Earth, and gave astronomers the chance to study its prominent dust trail. Comet Holmes, in November 2007 (referred to in the previous chapter), was even bigger: it had an icy nucleus of 1.9 miles across.[3] Detailed analysis of comets from the Stardust probe, which passed close to the comet Wild-2 some 242 million miles from Earth in January 2004, also revealed that the object was over 3 miles wide, putting it – like Comet Holmes – not just in the asteroid class, but in the planetoid class. It was covered, strangely, with craters, mesas and canyons, further blurring the distinction between comets and asteroids. In addition, a weird network of jets spewed material from the core of the comet into space.

In 1997 tens of millions of people saw the bright comet Hale-Bopp, one of the brightest objects seen for decades by people without optical aids. This comet was shrouded in a luminous haze – a startling sight. Yet it appeared not to move, only drifting away from its position very slowly over several days. At its peak it was travelling at over 70,000 mph, and its closest approach to Earth was a mere 125 million miles. Yet it was reckoned to be up to *60 miles across*! This belied, again, the usual assumption that comets are about the same as a giant boulder – say, about 50 yards across, the size of an average asteroid.[4]

When, in 2005, Nasa's Deep Impact spacecraft hurled a giant missile into the Tempel 1 comet to blast out a fluffy ball of powder to test its make-up, another space telescope monitoring the exercise detected clay and carbonates in the mix streaming off the comet. This raised a question mark over how clay and carbonates form in frozen comets. The satellite also revealed that a lot of frozen water had been ejected, more than expected. The outburst lasted twelve days and, in all, Tempel shed some 275,000 tons of water – enough to fill more than 260 Olympic-sized swimming pools.[5]

How many missiles are there in space?

There is still much speculation about how many celestial missiles are still up there in near space. There could be thousands of conventionally understood NEOs just waiting to crash-land on Earth, if one believes the list of likely objects drawn up by Dr Brian G. Marsden, of the Minor Planet Center at the Harvard Smithsonian Centre for Astrophysics. Other scientists said in 2009 that 'only a tiny fraction of asteroids less than 100 metres across have been catalogued ...'[6] The threat still involves the smaller rocks, according to a National Research Council (NRC) report released early in 2010. A 3-pound lump of rock was found near the alpine castle of Neuschwanstein in Bavaria, southern Germany, on 14 July 2002. This object was a fragment that broke away from the huge asteroid or comet that was up to 660 pounds in mass, about the size of a rock some 50 feet wide. Some 100,000 of the smaller asteroids and comets spanning 140 yards or more could deliver an impact energy of 300 megatons of TNT, they claim.[7]

UK astronomer Duncan Steel says that at least one dangerous-looking asteroid is being discovered now on a daily basis,[8] but usually when it is well into its Earth-bound trajectory. Astronomer Mark Bailey suggests that we probably see or experience only one comet out of every ten that exist. University of Arizona researcher Jay Melosh, and others using computers, reckon that a typical stony Near Earth Object (NEO) must be about 100 yards in diameter to make it through the atmosphere intact. Objects smaller than this, we are told, will break apart some 11 miles up, as did the 5-feet-wide stony asteroid that shattered into pieces and landed in the Chicago suburb of Park Forest on the night of 26 March 2003.[9] Before the impact there were said to be sizzling or crackling sounds.

Comets and meteorites are derived from planetary matter inside and outside the solar system. A meteorite, or *meteoroid*, is any piece of space rock smaller than an asteroid that is not completely burnt up and lands on Earth relatively intact (a meteor is an object that is still in space). The majority of asteroids are confined to the main asteroid belt, situated between Mars and Jupiter where a mythical and smashed planet was once said to exist. Asteroids are traditionally thought of as being stony, and rich in silicate minerals, with only small flecks of metal embedded in them. They can range in size from a small boulder to a large object several yards across weighing hundreds of tons, or even a mile-wide object capable of devastating huge areas of the Earth.

The risk of disaster to Earth inhabitants remains very real indeed, as we have seen from the fireball events discussed in Chapter 1. An 8-pound space rock crashed through the roof of a woman's home in Alabama in 1954.[10] The damaging effect of space fragments are a constant source of press stories. In Bogota, Colombia, four children died when their shanty house was struck by a fragment of asteroid rock.[11]

Missiles from space can reach such fantastic speeds when they finally penetrate the atmosphere – up to 25,000 mph and slowing near the surface to about 10,000 mph – that it means that their make-up is rather irrelevant: a ball of water about half a mile wide, and trapped within a thin balloon-like membrane crashing to Earth at that speed, would still cause a 1-kT explosion. Naturally, the smaller the space rock the greater is the probability of it entering Earth's atmosphere but, of course, the less damage it is likely to cause. We had one of our closest shaves in March 2004 when a 100-foot boulder flew past within 26,500 miles. This was a whisker away in astronomical terms, but could have caused immense damage if it had landed in the centre of a dense metropolitan area. Although the bolide would begin to disintegrate on its way down through the atmosphere, not enough would have been whittled away to prevent it doing harm. It could have killed thousands, according to Nasa.[12]

A crucial point: determining the orbits of asteroids and meteorites requires several weeks or even months of observation. Any new sighting would have to be assessed by other astronomers. But these experts cannot accurately work out what cosmic forces are at play that will enable them to say when we might expect a gravitationally tugged impact, or describe the destructive aftermath of that impact. Gravitational forces – certain in themselves and precisely calibrated – become vague in regard to their influence on the trajectory of bolides as they wing their way to Earth.

Orbital paths, in any event, are inherently uncertain. Indeed, every so often new and unknown missiles keep popping into astronomers' range of vision with barely a week or twos' notice. The sequences in the film *Deep Impact*, where a solitary one-off observation is fed by a schoolboy into a computer that instantly shows a hi-tech display of an imminent impact, is far from reality. For example, a fireball was watched by tens of thousands of witnesses across central Europe, but few astronomers noticed it at the time. The cameras on the European Fireball Network also recorded it after it had crashed on 6 April 2002. With little warning, both Hale-Bopp and McNaught, the 1997 comets, were discovered by astronomers who had given the world only two months' warning of their imminent arrival in our skies. Hale-Bopp's unforeseen and imminent arrival meant that no computer calculated, nor could it, its true path. And the predictions from the computers all differed about when the object would strike – varying from months to years.

Clearly, many giant natural objects from space do swing perilously close to the Earth, and can take astronomers by surprise. Further, some asteroids, having been spotted, can then get 'lost'. Such was the case of a half-mile-wide rock discovered in 1998, called OX4, which was tracked for only ten days before it vanished.[13] In 1997 an alarming report was published in *Nature*,[14] which said that another comet that passed by Earth would also, like Hale-Bopp, have been bright enough to be seen by the naked eye. Yet it was detected at the last minute by Finnish meteorologists using an instrument on the Solar and Heliospheric Observatory satellite. Teemu Makinen, one of the Finnish team, concluded that the failure to observe the comet (one of five, excluding Hale-Bopp) earlier 'underlines the need for full-sky surveillance of comets'.[15]

Difficulties seem to arise at the impact-assessment stage, where missiles can be ranked on a scale of one to ten in regard to how much damage they are likely to cause. But this 'damage ranking' can change over time once more observations have been made, and could drop easily from one to zero. A smaller asteroid 2004 MN4, said by Nasa initially to have been discovered in June 2004 and to be 400 yards across, was observed for only two days before disappearing. It was spotted again on 18 December 2004. At one time Nasa's website gave this object the highest ever odds of landing on Earth – more than one in fifty – then adjusted this later to a 1-in-300 chance of an impact in the year 2029. Jeff Larsen of the Spacewatch project at the University of Arizona said, when his team learned of MN4, that they disputed its trajectory, adding that the object had been detected in March 2004. After new data was input into the computers at the Jet Propulsion Laboratory at Los Angeles County, California, and the University of Pisa, Italy, scientists did some recalculations. It was then declared that the object would not hit Earth after all, although it would come as close as 18,000 miles, well within the outer orbit of commercial satellites.[16]

The defence implications of this type of explosive aerial activity are taken seriously. 'The fireball data from military or surveillance assets have been of critical importance for assessing the impact hazard', said David Morrison, an NEO scientist at Nasa's Ames Research Center, Moffat Field, California.[17] The link between nuclear tests and fireballs is still blurred, as we saw in Chapter 1, and a recent US military policy paper says that most cosmic missiles ought to be subject to security clearances before data is released. Critics say that this rule will not only hamper astronomical research, but the public understanding of sometimes dramatic sky explosions will be diminished. The fear of the unknown will increase.

This kind of speculation has goaded the CERN Particle Physics Astronomy Research Council to seek funding for a new robotic telescope to be set up in the Canary Islands. This could systematically scan the skies for meteorites, feeding data back to Armagh.[18] December 2005 was the new CERN goal to locate the smaller objects 140 yards or larger, to be finally totted up by 2020.

And it is not just natural objects that scientists are worried about. Nasa reports an unknown thing approaching Earth from deep space. Object 2010 KQ was detected by the Catalina Sky Survey in Arizona in early May 2010, and tracked by the NEO programme at the Jet Propulsion Laboratory in California. Its characteristics do not match any known asteroid types, and it is only a few yards wide. It made a close pass by Earth, coming almost as near as the moon. It is now headed away into space, but it seems to use no propulsion. This is strange, since an NEO on a gravity-tugged trajectory could only orbit a larger object and eventually crash-land. So this object must have moved under its own power at some point: 'One would not expect a naturally occurring object to remain in this type of orbit for very long', said Paul Chodas of JPL. The obvious conclusion was that it was a booster broken away from an alien spacecraft, 'since it is not one of ours'.[19]

Strange objects from the skies

Trying to account for the number and type of fiery rocks and boulders falling from the skies has been compounded during the last century by similar unexplained occurrences, usually described as Fortean phenomena. 'Fortean' is named after the early writer Charles Fort, the famous early twentieth-century collector of weird newspaper accounts of the paranormal and inexplicable, and who wrote several popular books on this theme. Shredded fish, tadpoles, tiny pebbles and odds and ends of all sorts have allegedly fallen to the surface of the Earth. Plastic buttons, for example, have dropped from the sky and blanketed several city blocks in postwar US cities, according to the late Brad Steiger, the writer on 'anomalist' events. At Bijori, India, beads of many colours, already holed and ready to string, have apparently fallen regularly for nearly a hundred years.[20]

There are other strange examples of falling rocks throughout history. The Abyssian army in the sixth century had to beat a retreat to a hail of stones, supposedly dropped by birds. At the time of the reign of King Tussis in Rome, 672–640 BC, a rain of pebbles fell around Mount Albanus.[21] A family of Sikhs in Kampala, Uganda was 'troubled and upset' by objects thrown.[22] A psychic researcher, Dr Nandor Fodor, detailed hundreds of examples of 'malignant spirits' that had thrown stones that had struck the walls of 'simple dwellings', dating back to AD 858. In Sumatra, Indonesia, a Mr W. G. Grottendieck awoke one night in 1903 to find stones raining down on him from his bedroom ceiling.[23]

In the early 1980s the residents of five houses in Thornton Road, Birmingham, spent several distressed years trying to ascertain why their quiet suburban street had been singled out for almost nightly bombardment of large stones that seemed to come from nowhere. Taking protective measures to repair windows and roof tiles that had been repeatedly damaged cost them dearly. The police were called in to keep the houses under surveillance, as it was assumed that mischievous vandals were at work. By the end of 1982 Chief Inspector Len Turley and his team had spent more than 3,500 man-hours in staking out the houses and gardens in fruitless investigations.[24]

Mrs Pauline Aguss, seventy-six, was hit by a cosmic missile in August 2004 when she was hanging out her washing in her garden in Lowestoft, Suffolk. She said the odd-shaped stone that her husband had found on a path 'looked black and peculiar', with small crystals embedded in its surface. In another case, a chunk of rock the size of a grapefruit crashed through the ceiling of a house in Auckland, New Zealand.[25] And in mid-April 2007 bright lights were seen over New South Wales which the authorities suggested were probably a break-up of a meteor. These lights were photographed, and shown at www.rense.com.[26]

In Marshall, Texas, on 11 July 1961, blazing red BOL nearly struck a man called Troy Peterson. The deputy sheriff said the smouldering object resembled a charred petrified lump of wood that sparkled and glinted when turned from side to side. It weighed some 16 pounds.[27] In June 1642 a lump of burning sulphur fell on to the roof of Loburg Castle, some 18 miles from Magdeburg, Germany.[28] An object from the sky made a hissing sound as it fell, and a shower of hot cinders landed in a V-shaped pattern about

500 feet from a farm in Ottowa, Illinois on 17 June 1857. The farmer said he noticed a small, dense, dark cloud over his garden before this happened.[29]

In Monroe, Connecticut, in the late 1970s residents thought an angry spirit was at work, because rocks were falling straight down for several days a week – usually in the evenings. Some crashed loudly on to roofs, and many were very warm, a common and inexplicable feature that people experiencing falls of stones or pebbles have noticed. However, after analysis they were found to be undistinguished local granites and rhyolites found in the area.[30]

In many cases falls of small fish or frogs from the sky were often attributed to 'whirlwinds'. The Greek historian Athenaeus refers to a three-day fall of fish and frogs which somehow had been sucked into the air. Sometimes these fall in phenomenal amounts: the people of Yoro, in Honduras, gather bucketfulls of fish, we are led to believe. Often they are caught alive.[31]

However, Charles Fort shrewdly pointed out that there is never any mention of other detritus, such as mud and vegetation from ponds, also falling along with the frogs. 'There is just', he wrote sceptically, 'a precise picking out of frogs ... a pond going up would be quite as interesting as frogs coming down ... it seems to me that anybody who had lost a pond would be heard from'.[32]

Falling ice has been attributed to fragments of ice comets, storm clouds, or lightning that fragments freezing rain particles which then accumulate in size. Often there is wild speculation whenever large blocks of ice fall from the sky, which investigative reporters do little to follow up properly. A rare ice fall of exceptionally large blocks occurred across Spain in January 2000. This stumped scientists because large ice-chunk sizes depend on how long they are suspended in freezing updrafts in cloudless skies.

Falls of ice are often blamed on the way accumulated ice becomes dislodged from aeroplane wings (although ice falls have been recorded before the advent of aircraft: the astronomer Camille Flammarion refers to chunks of ice falling at the time of Charlemagne in his *L'Atmosphere*, 1888),[33] and a massive 20-foot wide block of ice fell in 1849 on a farm on the Isle of Skye. The clear ice was composed of diamond-shaped crystals from 1 to 3 inches long.[34] A 22-pound block of ice crashed into a street in Yagotin, Ukraine, in July 1970, shattering upon impact into 'greenish crystals'.

Ice weighing some fifty pounds fell through a roof of a downtown Riverside official building in California, Los Angeles, in May 1972, and landed in a third-floor hallway. But the FAS said no aircraft was in the area at the time.[35] In another example, in July 2010, a massive block of ice left a hole in the ceiling of a home in Chichester, West Sussex, in the dead of night. Tiles and roof debris were scattered nearly 30 feet. The occupant and his wife airily asserted that the ice had fallen from a 'transatlantic plane', without mentioning whether Chichester was on a regular transatlantic flight path. A spokeswoman for the Civil Aviation Authority gave a garbled technical explanation: 'Aircraft have a number of seals, such as around doors. Ice does fall off sometimes when one of the seals fails ... or the pipes can drip but when you're flying at altitude that freezes. When you come into land it thaws and can fall off.'[36] However, a Spanish study by the Consejo Superior de Investigaciones Científicas (CSIC) said ice does not come from aircraft, as ice falls usually occur well away from airline flight paths.

Damage was done to a metal-clad awning overhanging a house in Washington DC on 5 February 1968, leaving a ripped-out panel. This was also said to be due to a chunk of falling ice, although the house was nowhere near an airport.[37] Several cars were damaged by falling ice along American Avenue in Long Beach, California on 4 June 1953. These were truly massive blocks, falling from the 'air full of white stuff'. Many of them hit a used car lot, and shattered into huge chunks. One witness said he heard a loud crash on his roof, and looking up could see 'the sun shining on big pieces coming from 2,000 feet up. They rolled and twisted and shimmered like some waterfall.'[38] A Spanish scientist, Jesus Martinez-Frias, identified some 200 ice falls as a larger species of hailstone, judging from the presence of air bubbles and layering in the 'ice'. The ice fall in August 2005 that landed through the roof of a home in Fontana, California, was described as a 'megacyrometeor', after Martinez-Frias coined the term.[39]

Other notable weird falls

1886 On 4 September in Charleston, South Carolina, stones bounced off the pavement outside the offices of a local newspaper.[40]

1922 In May, a chemist's shop in Johannesburg, South Africa, was bombarded with stones. The phenomenon was thought to be centred around a Hottentot girl employed there, and stones were seen by police to fall all around her. This went on, on and off, for weeks.[41]

1950 On 26 September two Philadelphia police patrolmen saw a 'shimmering' object drift down from the sky and land in a field. On foot they found the object, and described it as 'six feet in diameter and about one foot thick' at the centre. It appeared to be a purplish flying saucer in the light of their torches. When one of the sergeants picked it up it disintegrated, as if it were some kind of gelatine, leaving a thin, sticky, odourless scum. The entire mass evaporated within half an hour.[42]

1952 On 17 October a narrow cylinder was seen by many people over the skies of Oloron Ste Marie in France, accompanied by about thirty smaller objects. All were trailing 'angel hair' behind them. A large number of strands of this gossamer-like substance were found hanging from bushes and phone wires for several hours afterwards. It soon turned gelatinous and then disappeared.[43]

1961 A ton of polyethylene plastic film fell from the sky over Elizabethton, Tennessee, on 25 November, yet there was no plane in sight. Deputy sheriff Paul Nidiffer reported that the huge transparent sheet had no form to it, no starting or stopping place, and no identification tags.[44]

1962 A family home near San Diego during August experienced falls of rocks the size of walnuts, which came 'trickling down' from clear blue skies.[45]

1962 In September hundreds of limestone pebbles also came dribbling down over Clarence, Iowa, and some had specks of tar on them.[46]

1965 On 19 February in Bloomsburg, Pennsylvania, it rained tiny plastic pellets about the size of shirt buttons over an area four blocks wide and five blocks long.[47]

1969 A UFO seen on 20 March seemed to be a semi-circle of white light. It swelled and grew fainter while retaining a sharply defined globular area of light.[48]

1973 Two men fishing in a lake at Skaneateles, New York State, in October, saw one or two large stones splash into the water, followed later by smaller ones. The stones followed them when they went to their car. Geologists could only identify the stones as local rock types.[49]

1979   One Southampton witness said on 12 February that he often heard a 'whoosh' on the glass roof of his conservatory. He once found the entire roof covered with thousands of tiny mustard and cress seeds, coated in a kind of jelly. The day after this, a 'load of peas, maize and haricot beans' came down, while his neighbours on both sides were pelted with bean and pea seeds.[50]

1982   In September in Belize, Central America, a woman and her granddaughter were pelted with a continuous hail of small stones that seemed to come out of thin air while they were walking.[51]

Fireballs with 'tentacles'

There was growing alarm that many glowing objects seen in the skies did not 'stack up' as genuine space missiles, and there is no escaping the fact that the entire subject of 'fireballs' is wrapped up with the mystery regarding UFOs. Unfortunately, the word UFO is synonymous with, and interchangeable with, 'Unknown Aerial Phenomena' (UAP), and 'Lights in the Sky ('LITS'), and both these categories are themselves confused with space debris.

Despite the scientists' claim that they can distinguish asteroid blasts from nuclear blasts and from quakes, many remain sceptical. What often intrigued journalists, frightened the public and puzzled the astronomers, was the peculiar behaviour of bright careening bolides, fuelling the belief that these objects are indeed UFOs.

There are the oft-mentioned 'tentacles' of some flying objects: a series of trailing, often fiery, string-like appendages attached to the main fireball. 'Jellyfish'-type UFOs have often been reported, as also the 'mushroom effect', often seen on the underside of UFOs.[52] Astronomer Seth Shostak of the Cardiff Centre for Astrobiology conceded the truth of some of these apparitions. He referred to the fact that 'visible every now and then were long threads ... some of these were seen to reflect the light over a length of three or four yards'. He suggested they looked like spiders' threads, but very bright and that they reflected the sun.[53]

Sometimes seen are thin strands of gelatinous material, which float down from the sky and generally dissolve on contact with the soil, as we have seen above. A driver going north from Sacramento noticed 'a string of fibre-like material of various lengths, a few inches to many feet long'.[54] University of Wyoming microbiologists suggested these were genuine spider webs, due to seasonal migration of hatchling spiders.[55] They were sometimes said by scientists to be the product of the 'balloon spider'. But the quantities involved were way too high. During the night of 28 October 1988 coastguards patrolling the English Channel off Dorset reported a cobweb cloud with an estimated area of 30 miles.[56] On 21 December 2002 the *Galveston Daily News* reported 'Mysterious Webs Drop From the Skies': 'The skies literally filled with floating, shimmering strands and fuzzy luminescent wads that looked a lot like spider webs.' [57]

Dr Andrew Pike, the astronomer mentioned earlier who has long been associated with UFO studies (but generally as a critic), mentioned a 1930 copy of the *Marine Observer* in which a flash of lightning was seen entering the water in the Balboa Channel, Panama, some 100 yards off the port side of a ship. It seemed to be a 'streamer' rather than a 'ball of light' (see Chapter 6). It was slowly falling as if the lightning was dripping 'molten metal' (this was accompanied by the sound of broken crockery and 'sizzling').

In November 1975 members of a family in a sprawling suburb of Los Angeles saw a blue fireball, about 2 feet across, landing in a field nearby. 'The minute it hit, it let out tentacles of energy in all directions that covered the whole field', one witness said. A local paper said a fire started after the object shorted out an electrical transformer.[58] Sometimes 'fat contrails' are seen, with an odd description of 'dripping metal' that gets entangled with power lines. In the famous case at Rendlesham Forest in December 1980, a UAP was seen as being partly made of molten metal. This was dismissed by UFO critic Dr Pike, who blamed lightning-type electrical discharges.

On 12 July 1982 at Stamford, Connecticut, the electrical power slowly faded in a house while a bright orange and white object hovered about 50 feet high in the air outside. But it seemed to be floating above a nearby power company transformer, situated not far from the road. The witnesses said the glowing object seemed to be attached to the transformers by a glowing wire, perhaps using the power station as an energy lead. The United Illuminating Company, the local distributors, confirmed that a power drainage occurred in Stamford at that time, although the cause was unknown.[59]

An extraordinary fiery object was seen at Pencoed, near Bridgend in Wales. It was photographed with a digital camera on 24 September 2003 at 7p.m., and looked like an exploding fuel dump. But the resulting smoke trail behaved oddly and moved very rapidly across the skies. Nasa said it was the best picture of a meteorite they had seen, and made it into their Astronomy Picture of the Day. On its website, Nasa said it could have been a 'sofa-sized' rock that exploded as a fireball. Spokesman Robin Cathpole, a senior astronomer at the Royal Observatory, did however suggest it could really could have been an aeroplane contrail, a common feature attributed to glowing orange-red UAPs.[60] This theory was supported by the fact that there was no sonic boom, and there seemed a slight change of direction in the tail, which would normally rule out a meteorite, although a similar crooked fiery streak accompanied a Russian meteorite fall in the immediate postwar years. On 12 February 1947 a huge meteor was seen racing across the sky near the Sihote-Alin mountain range, halfway between Vladivostok and Khabarovsk. Later some 8,000 iron meteorites were found in the area, with the largest single fragment weighing 3,740 pounds.[61]

On 20 July 1964 several policemen observed a number of objects 'shaped like umbrellas' for several hours flying over Madras, Oregon. Sometimes they were stationary when not zooming off at fast speeds. They were reddish at first, but changed colour when they accelerated.[62] One witness noted that the objects were half-sphere shaped, with what he considered to be 'jets' and flames coming down to the ground from beneath them. One also gave the impression that a luminous cone beneath the half-sphere was helping to push the UAP higher. Then it shape-shifted, but appeared to be a dark circle with a luminous ring: it reminded him of a sunflower with yellow petals.[63]

The 'chemtrails' mystery

From the late 1990s, strange, low-altitude vapour trails in the sky have been alleged to be the work of aircraft spraying harmful chemicals into the air for various nefarious purposes, such as the testing of toxic weapons, or perhaps weather modification experiments are being attempted.[64]

There was a discernible difference between quickly evaporating contrails and the longer-lasting, often tinted trails speading out and criss-crossing the sky in streaks. In Parsonsfield, Maine, from the late 1990s through to the early 2000s, a couple noticed chemtrails remaining in the sky longer than normal contrails did, and that they often ran side by side in parallel lines. They also appeared far from established air routes. One witness in Arizona said he could see normal contrails higher up, but the other chemically induced contrails were 'broad, low-hanging contrails running at angles mainly east-west across the sky and spreading....[65]

Atmospheric researchers working for aircraft companies have revealed that there are pollutants in the chemtrails, and this has been confirmed by LIDAR (Light Detection And Ranging). One motorist said that on the Interstate 65 highway from Indianapolis towards Camp Atterbury he counted 27 separate chemtrails. Another said the bands were up to 40 miles wide, over Crawford in Texas, where some eighteen jet aircraft were flying in formation.[66]

In his book *Chemtrails Confirmed* (2004), William Thomas chronicles at length the various flu-like illnesses people in the USA have experienced. Hospitals in southern California and in Oregon had emergency rooms with people suffering from respiratory illnesses.[67] According to the internet, people are still suffering from respiratory diseases attributed to chemtrails in the late 2000s. It was alleged that the planes had USAF markings (silver or white with red bands). Planes were said, in one particular instance, to fly back and forth over Yuma, Arizona, making the sky 'solidly overcast'. Aeroplanes were seen laying chemical trails in a grid pattern over New Mexico where they created artificial cirrus cloud layers which soon turned to hazes. Many people said they suffered from headaches, nosebleeds, hacking coughs, etc.

Oddly, the spider-web-like substances have often been attributed to UFOs, and the term 'angel hair' has been used both by UFO witnesses and chemtrail witnesses, as we have seen above. String-like or gelatinous substances from chemtrails also bizarrely emulated what some space objects were doing. 'Spider-silk' droppings apparently came from covert plane exercises, presumably operated by USAF or official agencies. In Sallisaw, Oklahoma, a witness saw 'cobwebbing stuff coming down' from zigzagging jets flying 'all day long, line after line ...' [68] A doctor from Sedona in California said he observed 'hundreds, perhaps thousands' of 'filaments' seemingly drifting in the wild. 'Angel hair' was observed in Tennessee, New Jersey, Ohio, Colorado, Oklahoma and Oregon. On rigs in the Gulf of Mexico an oil worker was stunned to see a 'white web or angel hair type stuff'.[69] In the Californian mountains, near Yosemite, people saw the web-like subtances descending from a height of 8,000 feet – 'long strands of it, 20 foot long, for hundreds of feet'.

Examples of 'molten' or gelatinous falls

1652   A meteor landed and left a 'star-jelly' substance on the ground.[70]

c.1818   A fiery globe fell on the island of Lelthy, India, and left gelatinous substances.[71]

1833   In November in Rahway, New Jersey, a jelly-like substance was left behind after a 'fiery rain'. It disintegrated into a white powder, before finally disappearing.[72]

1835   In August, a shower of stones appeared from a small dark cloud, possibly tinted yellow or red, at Marsala on the western coast of Sicily.[73]

1846   In November, a 4-foot-wide 'luminous object' left behind a 'foul smelling' jelly in Loweville, New York.[74]

1958   A BOL landed in a field at Westmeath, Ireland, in February. It left gelatinous material on the ground.[75]

1962   A fisherman in Concord, North Carolina, in September saw a glittering bowling-ball type of object fall unnaturally slowly to the water's surface. It was studded with short spikes, rather like an early communication satellite. By the time marine officials had arrived, it had begun to disintegrate, and became a mass of shiny wire or shredded aluminium. It soon dissolved.[76]

1981   In December, at Greenwich in Connecticut, a 'silver ball' tailed a car and then shot down three laser-like beams of red light that headed towards the boot of the car. These left silver spots of damage on the car's bodywork.[77]

2004   A 'jellyfish' type of UFO was seen on 14 April at Honley, West Yorkshire, with two bright lights stuck on either side. A Birmingham man saw an illuminated blue triangle hovering over his garden which then took off, leaving a 'silky-white' substance on the tree tops.[78]

2004   Contrail lights, looking like a string of something bursting into four orbs and then taking off in a boomerang shape, were seen in broad daylight at Phoenix, Arizona, on 30 October. This and other worm-like or string-like coloured light entanglements were shown in video footage reproduced in *UFO Data* magazine.[79]

2009   The clearest 'sausage' contrail was shown on the *Daily Telegraph* website of April. It was photographed at night on 31 March 2009 at South Harrow, north-west London, and showed light traces to the left. It was stationary, then moved away at a rapid clip. It made a 'whirling' sound.[80]

The molten characteristics of some phenomena were often connected with electrical 'sizzling' sounds, and some UFOs were seen hovering near electrical sources. In the 1975 Los Angeles example, mentioned earlier, we saw that an object shorted out an electrical transformer. The Stamford, Connecticutt event of 1982 – where a glowing object attached itself to a power station with some sort of lead – is also curious.

In a later chapter I suggest that UFOs, as sentient flying objects, actually seem to glean power from the cosmic environment. But many, perhaps, need to zap energy from human sources. Some witnesses say UFOs are seen near power stations and that their activities are responsible for occasional blackouts.

But, of course, there are other more mundane explanations, as we shall see in the next chapter.

3 The Blackout Mysteries

Some fifty million North Americans found themselves plunged into darkness on the night of 16 August 2003, when the electricity power failed. Those most affected lived in the north-east, primarily Ontario, and six states in the eastern regions of the USA. But very little light – in a manner of speaking – has been shed on the cause of North America's biggest blackout ever.

It was suspected that these power failures, which occurred also in Europe, were the result of the extraordinarily severe solar flares of that year. Flares, particles and cosmic rays are part of the phenomenon known as 'space weather', and are reckoned to arise from physical processes that are quite distinct from terrestrial weather. Ordinary weather results from dense, electrically neutral gases in the Earth's atmosphere, reacting in complex ways, whereas 'space weather' follows laws that govern *plasma*. Earth is situated in a region of space that is loaded with this esoteric gas, which is essentially an electrically charged vapour consisting of tens of billions of shattered atoms, where the gases and dust of space have had their internal particles broken down, and where the atoms lose their outer electrons.

A spokesman for FirstEnergy Corp, Todd Schneider, said the utility company's data showed unusual fluctuations of voltage in the Midwest grid, more than three hours before the company's lines failed.[1] The technical faults were said to be located somewhere on what is called the Lake Erie Loop. A sudden reduction in power along the grid seemed to change the flow on the northern leg of the loop, when 300 megawatts of electricity reversed direction from east to west. This caused the generators and other systems to 'trip' across the region, a shutting-down process designed to protect the equipment.

The radiation spewed out by the sun continued from the summer well into the autumn. Two further giant solar fireballs that hurtled to Earth in late October 2003 were widely reported in the world's media, and it was unusual to have two flares so close together. The official reticence about the 2003 solar flares affecting Earth's electricity grids could be explained by Nasa's and the Department of Defense's fear about its spy satellites, according to sunspot-watcher Mitch Battros of the organization Earth Changes TV (ECTV). Perhaps as many as forty-five Coronal Mass Ejections (CMEs) emerged from the two major summer flare events, according to Battros.[2] In the name of 'national security' these organizations could not confirm or deny any damaging solar events.

This reticence might also have been due to a large cylinder-shaped object that was seen and recorded on video film near Fostoria, Ohio, only 30 miles from Lake Erie, and also over New Jersey.[3] The *Niagara Falls Gazette* featured a report about dozens of people witnessing UFOs hovering over the power plants near the Falls.[4] It was these later UAPs that belatedly got the attention of various public and scientific bodies such as the National Oceanic and Atmospheric Administration (NOAA) and Nasa. However, the DoD decided to 'go public' about the solar connection, but not about the UFOs.

Solar flares and power failures

Scientists realize that powerful electrical charges that are zooming around in near space are having dramatic – even disastrous – effects on humans living here on the surface of the Earth. When any solar flare or CME penetrates into Earth's atmosphere, it has an immediate impact upon our lives. The complicated systems that support us, such as water and sewage treatment, supermarket deliveries, power station controls, and gas and fuel pipelines, are all vulnerable. Most of our electrical infrastructure, including mobile phone networks, household appliances and the internet, would stop working. Ordinary vehicles are also at risk.

A CME that happened on 14 July 2000, and which was detected at the Space Environment Center in Boulder from signals from the GOES-8 satellite, revealed a sharp jump in the intensity of solar

X-rays. These seemed to come from a highly active solar region that had been roiling for that week. Domestic lights would brighten before going out. 'Millions could die' we were warned in January 2009 from a joint Nasa/NAS report, referring to both the 2000 and 2003 events, and published in the *New Scientist*. The Nasa report reminded us that electricity grids are nowadays designed to operate at ever higher voltages over larger areas.

'We are moving closer and closer to the edge of possible disaster', said Daniel Baker, a space weather expert based at the University of Colorado in Boulder, and chairman of the NAS committee.[5] A vast cloud of gas that had a temperature of 1.8 million degrees F, and was 'more powerful than a million hydrogen bombs', according to Nasa scientists, struck Earth's atmosphere on 28 October 2003, impacting upon the Earth's geomagnetic field.[6]

The erratic nature of exploding flares often leaves researchers with only thirty minutes' warning, and as little as seventeen hours to prepare for speeding clouds of plasma. CMEs have temperatures of more than one million degrees F, according to Don Smart of the USAF Philips Lab in New Mexico. The largest, he says, carry energy enough to boil a lake several thousands of times the size of the Caspian Sea.[7]

A failure of transmission lines occurred at Itaipu, the world's largest operating hydroelectric plant across the Brazil–Paraguay border, shortly after the Nasa report was published in the Western press. This created a domino effect that cut energy to large swathes of Brazil and Paraguay for more than two hours in November, and affected sixty million people. This was alleged to have been caused by an 'unexplained atmospheric event'.[8]

In the meantime, Mitch Battros is nervous about sunspot region 380: 'It is very large and is set dead centre, which could produce a direct hit on Earth ... if Earth experiences a direct hit from any one of these M-class or X-class flares, it could cripple our infrastructure.'[9] He said: 'Watch for freak storms to continue, while tornadoes or tornado-like winds, sudden rain, hailstorms and record-breaking temperatures are likely.' We could have only a ninety-second warning, according to Nasa's scientists in regard to any newly discovered solar flare.

There is another fear about rogue electric currents from outer space. The 'solar wind' consists of hot charged particles streaming

outward like a wind from the sun. It blows over our planet like a magnetic hurricane. Most of the solar wind, fortunately, is deflected by Earth's magnetic field, which stretches 40,000 miles towards the sun, and normally protects us from damaging cosmic and solar particles. But when a reversal of the field occurs it could have untold and even unknown consequences. And there is already evidence that Earth's field is 'on the turn'. Nils Olsen, of the Centre for Planetary Science, Copenhagen, agreed with the possibility of an imminent magnetic flip. He said the Earth's core could be undergoing dramatic changes. He said: 'This could be the state in which the Earth's geodynamo operates, before reversing.'[10] The last reversal was 250,000 years ago, according to Dr Alan Thomson of the British Geological Society. He pointed out that as there has not been a reversal for almost one million years we are due for one now.[11]

Indeed, a decrease in the power of Earth's field began some 2,000 years ago, but sped up rapidly over the last twenty years, and has become erratic, according to Nasa, referring to the THEMIS spacecraft readings. Alarmingly, there is an apparent breach in Earth's field ten times larger than before. This could cause seriously disruptive geomagnetic storms to reach the surface. The discovery came on 3 June 2007, when five probes flew through the breach just as it was opening, and a 'torrent' of solar wind particles was recorded by sensors. David Sibeck, the THEMIS project scientist at Goddard Space Flight Center, Maryland, said: 'This finding fundamentally alters our understanding of the solar wind-magnetosphere interaction.' [12]

'Fireballs' over power stations

Important geological and geographical factors, as well as solar factors, play a crucial role in these damaging power station failures that have recurred at other times. On 17 August 1959 Brazilian power stations were cut off in the Minais Gerais region, and UFOs were supposedly seen at the time.[13] On 14 February 1963 there was a sudden blackout in Denver, Colorado. Two linked power stations suddenly stopped functioning. This was attributed to an overload on the lines, and on the relays to Cheyenne and Boulder. But the power cut ended inexplicably and suddenly.[14]

Earlier, the chief engineer for New York's electrical utility, Consolidated Edison, after studying power failures of the 1950s, became convinced of the solar connection. He was aware of the 1959 Brazilian power failure, and he also noted a New York power failure of 17 August of the same year. They both caused interferences with shortwave radio transmissions. On his advice, Con Edison dug up the streets of New York and placed special shielding over the city's lengthy network of underground cables. This move dramatically reduced the number of power failures. But other power companies responsible for the huge north-eastern power network didn't follow suit.

Still another blackout occurred in November 1965. As with the 2003 flares, there were reports of 'fireballs', and it is alleged that pilots saw a large red ball of light about 100 feet in diameter hanging in the sky directly over two 345,000-volt power lines at the New York Power Authority's Clay substation. 'Lights in the Sky' (LITS) were also seen over Syracuse, New York State, and over power lines; they were photographed, and published in *Time* magazine. These fireballs, somewhat inevitably, were dismissed as a cause of the blackout.

Public figures became concerned about the 1965 power failure, which may have been due to the heavy consumer demand for electricity on a cold November night, with a power surge along tripped lines surging along other lines, overloading and tripping them in the process. Some thirty million people were affected, and thousands of square miles of territory were plunged into darkness. The overload automatically closed down substations from Ontario, New York, New England and Maine.

But the reports of UFOs didn't help matters. Several editorials called for an official investigation, and the major television networks took an interest. Some twenty-one of the nuclear monitoring locations had mysteriously been blanked out, while other sensors located in Salt Lake City in Utah, and Charlotte in North Carolina, flashed red signals indicating that nuclear explosions had occurred. A secret government establishment based in Virginia known as 'Mount Weather', said to be an underground nuclear shelter for the political and military elite in the 1960s and 1970s, was under the control of the Office of Emergency Preparedness, later the FEMA. This shelter closely resembled a fully equipped

Cold War underground city, with hospitals, office buildings, roads, sidewalks and radio and CCTV studios.

Another power failure occurred on 4 June 1967 that blanketed out much of the north-eastern USA, affecting mainly Pennsylvania. Again LITS had been seen beforehand.[15] Both the 1965 and 1967 power failures were said to have been triggered by an incorrectly set protective relay on the transmission lines at the Niagara generating station in Queensland, Ontario.

Yet another power failure occurred in New York in July 1977, which happened without warning on a typically stifling and humid evening. This time there was looting and disorder, resulting in 2,500 being arrested. Westchester County, north of New York, was struck by lightning, with the ensuing fire causing an explosion in a nearby transformer, sending flames 300 feet into the air. Transmission lines linking upstate coal-fired power plants with the city were severed. A nuclear power plant half a mile from the damaged transformer was immediately shut down on safety grounds.

Quebec, in March 1989, suffered a massive power supply failure that created problems on a 345-kV line near Cleveland, Ohio, which triggered a cascade of other problems that affected huge areas of the Canadian province for four days and inconvenienced some six million people.

Again, in 1995, the USA's largest airports experienced unprecedented power failures, which affected New York, Chicago, Washington DC and Miami, where seven outages were recorded during a three-week period. In January 1996 more power cuts were experienced, interfering with Seattle Center's ability to contact some fifty planes flying over a 286,000 square mile area, which included Washington State, Idaho, Oregon and Montana.[16]

Canada, in recent years, has been investing heavily in computer technology – more so in fact because its thirty-three million population is spread out over an enormous geographical area and so depends on electronic forms of communication, and the greater the distance the current has to travel, the greater is the voltage difference induced by the magnetic fluctuations, and hence the stronger the current that is needed. Canada's high-tech infrastructure supplies phone and television links all the way from Labrador to Vancouver. Although the additional voltage differences add barely a few hundred kilowatts to the tens of thousands already coursing along the grid, they can still upset delicately calibrated voltage regulators so that circuit breakers become disabled. Recall also that Canada is near the North Pole, where intense magnetic fields exist, and it is significant that Sweden also experienced power cuts in 1989 (in poor conducting regions located on a shield of igneous rock, the currents gladly surge through the power lines).

The problem with LITS that so frequently accompany power station failures is the difficulty in trying to sort out the natural from the unnatural, and from factors inherent with highly charged electrical grids themselves. A predicted solar burst for September 2012 will produce pillars of 'incandescent green light writhing like gigantic serpents across the skies', according to science writer Michael Hanlon.[17]

UFO-type car ignition failures also seemed to occur at the time of grid failures. Electrical power failures led to a garage owner claiming to have dealt with over thirty cars in the four months of November 2005 to February 2006. The area affected was near the Type-93 radar dome at Trimingham in Norfolk. It was suggested that a faulty MoD radar zapped unsuspecting motorists and that this caused engine failures. (East Anglia has a long tradition of UFOs arriving from over the North Sea, and eyewitnesses say they are often seen near the Trimingham radar site.)[18]

Scientists at the Spacecraft Environment and Protection Department at Qinetec, Britain's privatized defence research institution and arms dealer, had already expressed concern about the safety of satellites.[19] And for good reason, because the Russian craft *Soyuz XI* lost contact with mission control as it passed through the ionosphere in June 1971 after 124 days in space, but all three cosmonauts were later found to be dead. A Turkish scientist claimed that the cosmonauts' deaths were due to atmospheric electricity that brought about a condition known as alkalosis, where a pathologically high alkali content in blood and tissues leads to respiratory failure. [20] In the summer of 2004 a huge power failure left half of Greece, an area with more than five million people, without electricity for several hours, just prior to the Olympic Games. Officials at DEH, a Greek state-run power corporation, scrambled to find the cause. Suspicions about a solar connection clashed with other exotic theories, especially when reports of a 'very loud explosion' at a generating plant at Lavrion, 50 miles south-east of Athens, were mentioned in the media. Temperatures were close to 104 degrees F in July, so there was no power for air conditioners, and hundreds of commuters were stuck in Metro trains and central Athens became gridlocked. The blow-out affected the south of Greece, and all the Aegean islands were left without electricity.[21]

Some of these exotic events, on closer examination, were traced to occurrences much closer to the surface than could have come from solar flares emanating from ninety-three million miles away. Too often communities were left without electricity for no known reason. Paradoxically, in some cases this occurred while the atmosphere was dramatically illustrating just how much latent electricity there is in the sky. This celestial power does everything that bolides and solar flares are also said to do – create bright flashes in the sky with an accompanying explosion.

How powerful and how destructive these sky explosions are during lightning strikes we shall examine in the next chapter.

4 Electrical Storm Lights

The most dramatic and recognizable form of electrical illumination in the skies is, of course, *lightning*. But assessing the number of lightning strikes around the world is an inexact science, largely because of the looseness of the scientific definitions. 'Lightning bolts', 'thunderstorms' and 'lightning flashes' are all common and interchangeable expressions. *New Scientist* once estimated that 1.2 billion lightning *flashes* occur every year, mostly in the tropics and the sub-tropics, in thousands of thunderstorms around the world. Physicist Louis Proud reduces this figure to about sixteen million lightning storms occuring around the world annually.[1] A typical tropical storm has the energy of 10,000 one-megaton hydrogen bombs, according to one scientist.[2]

Other estimates of the likelihood of being struck range from one in five million to one in 600,000.[3] TORRO, the Tornado and Storm Research Organization, says there are around 300,000 lightning *strikes* to the ground each year in Britain alone, and about thirty to sixty people are struck each year, with around three people killed.[4] In the USA, annually, the number of people injured is between 200 and 1,000 – and four out of five are men.

Some kind of runaway atmospheric breakdown takes place to cause these lightning strikes, with massive air movements clashing at different temperatures. Meteorologists believe that fundamental positive-negative interactions apply to the atmosphere, which charges itself up to millions of kilovolts just through kinetic energy arising from the dynamics of a moisture-laden weather front. Moisture is drained from milder air, which cools into a mass of heavier cold air that then rockets downwards, and fans out in all directions. The upwards and downwards kinetic movement of layers of dense atmosphere can carry small water droplets to heights between 35,000 and 70,000 feet.

This strong electrical coupling between the various layers of the atmosphere causes thunderstorms. The charge in the moist air gets bigger until it overcomes the normal restraints of the atmosphere's insulating properties, and a current of electricity forces a path through the air to make its violent connection. A study at the Kennedy Space Center in Florida showed that about a third of all such strikes can travel 5 miles horizontally from their point of origin in the air.

Negative charges always seem to be at the base of the cloud, with the positive at the top. The negative charge in the cloud shoots down to the ground in a forked pattern via a channel of *positive* charges reaching up from the ground, usually through something like a tree or a telephone pole, and this completes a circuit. So the visible lightning is actually one or more return strikes as the current makes its way back to the cloud.[5] Ron Holle, a meteorologist who studies lightning for the Global Atmospherics Institute in Tucson, Arizona, said some lightning strikes could instead be one occasional flash with multiple return strokes: 'Between the strokes there is a continuing current, and it doesn't stop. We have no idea why it happens.'[6]

Charlie Moore and colleagues at the New Mexico Institute of Mining & Technology in Socorro, New Mexico, found X-rays coming from a 'leader' of a nearby lightning bolt. This is like a current that judders its way down to earth via a pathway. The positively charged lightning, according to this reasoning, rebounding from the ground, is much more powerful than negative lightning already in the clouds. The positive charges can carry up to ten times the current, and can reach up to 300,000 amps and one billion volts.[7] Some regions seem high in voltage, like the Great Plains of the USA, where storms can generate electrical strikes downwards from an elevation as high as six miles and lightning strikes can cover areas up to 100,000 square yards.

How powerful are lightning bolts?

However, some scientists like Greg Leyh, an electrical engineer at Stanford Linear Accelerator Center (SLAC) in California, consider this 'billion-volt' analysis to be wrong. Leyh argues that the Earth-cloud syndrome is surely not powerful enough to trigger lightning. He argues that the clash of electrical polar opposites between clouds and between Earth and the clouds could not be satisfactorily explained in the absence of Earth's electrical field. He doesn't believe that lightning can reach anywhere near one billion electron-volts, and if it did it would likely create instabilities in the geomagnetic field and short circuit part of it. Thomas Marshall of the University of Mississippi agrees that there would be no particles with the right kind of energy to trigger off all Earth's lightning bolts.[8] Experiments using kits and balloons show that the electrical fields rarely top 200 kV per metre. So how does the sky arc with its 'streamers' when allegedly the voltage is so low?

Indeed, there are some strange anomalies that seem to have no connection with the ionosphere, the geomagnetic field or a moisture-laden sky. Cosmic rays could hit the atmosphere, ionize a chain reaction, and the resulting cascade of electrons could trigger a lightning strike even in modest storm fields, according to Thomas Marshall of the University of Mississippi.[9] In June 2007, according to the *Miami Herald*, a Florida man was killed by a lightning strike that seemed to come out of a clear blue sky. This in itself is surprising because even the cosmic rays by themselves may not be energetic enough to do this, and researchers are still trying with computers to ascertain whether lightning strikes coincide with cosmic ray showers.[10] Scientists at the Lebedev Physical Institute in Moscow say that it could be the more powerful gamma rays that could do the trick. Here, then, is evidence of a link between all kinds of LITS – electrical discharges – many emanating from *beyond* the atmosphere.

Freak aurora lightning

New data from Nasa satellites shows that the *aurora borealis*, those stunning curtains of translucent colour that appear high up in the northern skies, are formed when solar particles are whisked along temporary 'magnetic ropes' connecting the sun to the Earth in what is called a *substorm*. These 'ropes' linking Earth's upper atmosphere directly to the sun are favoured in the springtime when the Earth's field is in its best orientation. This opens the door for solar wind energy to flow in. In doing so, it heats the gases in the atmosphere, and make the electrons in the air's molecules move into the visible spectrum. The light particles that glow with characteristic spectrum colours, notably red and green, result from the energized particles in the oxygen atom.

It is surprising to learn that the aurora borealis, normally visible only in the highest latitudes, can sometimes be seen in the UK. The astronomer Edmund Halley saw them from London in 1716. Auroral-type lights were also observed at the time of the giant solar storm of October 2003, when massive heatwaves occurred across Europe and resulted in giant power failures. A green glow was reported at 6 p.m. on Wednesday, 29 October, in that year in Ireland. Mike Tullet, a retired meteorology lecturer, said: 'It was spectacular, as if streaks of light came down to the ground with the colours of the rainbow, rather like shimmering curtains; it was so bright that at times the stars were obliterated.' [11] An observer in Northumberland reported that the sky 'turned to green in waves' and in the New Forest, in Hampshire, green bands appeared 'like a barcode'. They were seen over Cornwall, Sussex and as far north as Scotland, and lasted months.[12] The Nasa satellites also monitored an aurora display in March 2007: 'The auroras surged westward twice as fast as anyone thought possible, crossing fifteen degrees of longitude in less than one minute. The storm traversed an entire polar time zone, or four hundred miles, in sixty seconds flat', according to Dr Vassilis Angelopoulos, of the University of California at Los Angeles. The two-hour storm released 50,000 billion joules of energy – the same magnitude as a 5.5 earthquake.[13]

In the meantime, a weird and dramatic form of lightning, millions of times more powerful than the lightning generated in weather storms, has recently gained the attention of scientists. Pilots have also reported, over many decades, glowing objects that loom way above the clouds. The Apollo mission astronauts regularly reported seeing ghostly flashes, streaks, 'spots' and 'clouds of light' in space, and many of these are alleged to have been UFOs.

Over the past decade, electronic instruments have recorded two main categories, 'red sprites' and 'blue jets' that burst upwards from cloud tops to a very great height, with the blue jets shaped like water fountains.[14] The sprites are usually shaped like a carrot, turnip or jellyfish, or are 'dancing red blobs' with tentacles. Other more unusual sprites are described as pancake-shaped, seeming to be at a very high altitude of up to 40 miles, and have a reddish glow.

The sightings at first were attributed to a strange quirk involving the pilots' and astronauts' colour and light sensors at the back of the retina (see Chapter 8). Perhaps it was an afterglow affecting the pilots' perceptions of something that had not been directly observed. Although they are seen in just a fraction of a second, they can now be recorded on sensitive electronic equipment like scintillation detectors. Scientists now suspect that the sprites are not so high up in space, since they often manifest themselves lower down near the surface, and their microsecond visibility window can be larger and longer than usually thought. Perhaps heavier cosmic particles, such as helium and lithium, are responsible for the flashes, or perhaps they absorb and magnify electrical energy from terrestrial storms. The space shuttle video imagery hints, incidentally, that the flashes are the result of a strong relationship with ordinary lower-down lightning strikes.[15] The lights give off ELF radio waves – a sign that they are electrical discharges between the ionosphere and the clouds below. (An example can be seen at http://quicktime.oit.duke.edu/news.sprites.mp4.)[16]

Lightning damage to infrastructure

Lightning can be dangerous not so much because of the total kilovolt energy that is gleaned from the skies so much as the vulnerability of our built-up environment, and that of our defence and security infrastructure. If a lightning bolt strikes the ground, its intense heat can melt and fuse sand into a glass known as fulgurite. Entire houses can be blown up or set on fire, with brick walls and roof tiles providing no protection. Lightning had once made the walls of a house too hot to the touch for a fireman in a 1961 house fire in Alstead, New Hampshire. This could not be explained or understood, since there had been no fire, and there was no blown fuse or even a short circuit.[17]

Lightning discharges have had all kinds of sounds associated with them: smashing crockery, sizzling fat, sonic booms, splintering wood and the sound of footsteps.

In olden days church vaults and castle keeps were often used to store quantities of gunpowder. This clearly was a dangerous practice because the church steeples readily acted as lightning conductors. On 27 October 1697 a fierce thunderstorm erupted over Athlone, in Ireland, destroying the castle and many surrounding buildings. There were reports of a fireball; historical archives spoke of a 'wonderful and great body of fire ... which fell down the magazine, took fire, and blew up the granadoes [hand grenades]'.[18] It also blew up 260 barrels of gunpowder, plus hundreds of pieces of other types of explosive and kinetic ammunition, such as musket balls. The town had to be virtually rebuilt.

In 1769 lightning hit the tower of a church in Brescia, near Milan, and blew up the 100 tons of gunpowder stored there. One-sixth of the city was levelled in the explosion which killed some 3,000 – possibly the world's worst recorded lightning disaster. In 1807 lightning struck a gunpowder warehouse in Luxembourg City. The resulting explosion destroyed buildings all around, and killed 230 people.[19] York Minster's south transept was set ablaze by lighting on 2 July 1984, which some claimed was divine retribution after the Bishop of Durham, the Right Rev David Jenkins, made controversial remarks about Christianity.

Metallic structures such as railings, radiators, structural girders, iron conduits and even car bodywork naturally tend to increase the receptiveness of lightning-hit areas. Low-frequency fields can become localized near plumbing systems with grounding connections to metallic water pipes and other electrically conductive sources that are left exposed and vulnerable. Lightning does occasionally strike indoors through outdoor metal pipes and cables.

Selected examples of lightning dangers

1994 Britain's worst lightning strike was responsible for an industrial disaster when an oil refinery was hit at Milford Haven, Pembrokeshire, on 24 July. A valve was struck by lightning, which led to a pipe rupture, and 20 tons of liquid hydrocarbon exploded with a force equivalent to 4 tons of high explosive.[20]

2001 A Kent woman was hurled out of her bath when a bolt of lightning struck an outdoor water pipe and surged through the plumbing into the showerhead she was holding.[21]

2003   A golfer was struck twice in thirty minutes through the metal spokes of his umbrella, and another was killed sheltering beneath a tree.[22]

2006   In Colombia on 20 November, lightning hit 28 people and killed five of them while they were watching a football match.[23]

2006   The thunderstorms and lightning strikes in August that hit Britain all the way from Ross on Wye to Glasgow 'had impacts that sounded like bombs'. Some houses caught fire, and many roofs were damaged. In Kidderminster, Worcestershire, a mother and her baby were hurled 10 feet across a bedroom when lightning struck their house. And an eight-year-old girl narrowly missed being hit by lightning when a bolt blasted through the roof and set fire to her bed in Fernhill Heath, in Worcestershire.[24]

2006   In August, a pensioner was caught up in a blaze when lightning struck her home in Bromsgrove, also in Worcestershire.[25]

2006   A Wiltshire schoolgirl was hit by lightning and had burns to the hands when a bolt hit an aerial on the roof, blew a hole in the bedroom ceiling, and struck her metal-framed bed.[26]

2006   In October, a woman in Croatia was in the bathroom when lightning struck a water pipe outside. She was later told, after recovering consciousness, that the lightning bolt had travelled down the pipe and struck her on the mouth, passing through her body.[27]

2006   In Tuscany, in Italy, during the summer, a man who was standing by the metal steps of a swimming pool was killed by lightning. The coroner, in Exeter later, said it was one of the most bizarre deaths that he had dealt with.[28]

2008   In November, a bolt of lightning struck a fireworks warehouse near Sivakasi, southern India, triggering explosions that destroyed the building and about forty houses, and caused multiple deaths.[29]

Lightning has been blamed for aircraft crashes during thunderstorms, although this doesn't often happen. On average, lightning strikes a commercial jet once every 10,000 hours it spends in the air. Sometimes, however, severe bolts can cause tiny holes to appear in the fuselage and wings. A 'microburst' struck a Delta Airlines jumbo jet Lockheed L-1011 in August 1985 when it was approaching the Dallas-Fort Worth airport. It crashed far short of the runway and careened into an airport storage tank.

Generally, though, planes and cars are safe places to be. In fact, the surrounding tube-like container can act as a 'Faraday cage', dissipating the electrical charge harmlessly by conducting the electrical current across the surface of the metal fuselage. Even in rare cases when a plane is damaged, it can still fly. In April 1977 lightning ripped the nosecone off a private jet, but it landed safely. Rockets are safe too, as launch vehicles from Cape Canaveral have been hit several times by lightning without mishap.

Even so, the computer software of modern aircraft can be a natural hazard in itself. The danger of electromagnetic (EM) signals disrupting the circuitry of computers and electrical appliances is now a very real one. Meteorologist John McCarthy tried to warn the US Federal Aviation Administration about the threat.[30] Cockpit instruments, especially, are too often vulnerable to massive EM surges, which can result in false readings. When lightning hits an aircraft in flight it could immobilize the artificial-horizon display instrument panel, and cause the aircraft to crash, as possibly happened to a commuter aircraft on 24 May 1995 flying through fog and torrential rain across Yorkshire, killing all twelve people on board.[31] A Russian airliner crashed and burst into flames in eastern Ukraine during a storm on 22 August 2006, killing all 170 on board (the plane was a Tupolev 154, less vulnerable than most Russian aircraft). Further, some airliners have devices which are more sensitive than others. Fighter aircraft rely on computers to control the engine and aerodynamics. But these must work alongside high-power communication transmitters and devices that emit powerful jamming signals. There is an occasional loss of communication, which usually occurs in cloud, and a failure with positioning instruments can result in pilots becoming disorientated.

Even without lightning strikes, strong electronic or EM signals can cause planes to crash. This appeared to have happened to an Air France passenger flight from Brazil to France in June 2009, when all 234 passengers and crew died. In 1984 a West German air force Tornado crashed, killing its crew, after flying too close to a group of radio transmitters near Munich. In the late 1990s the air exclusion zone for Fylingdales, the signals reception and aircraft warning centre in Yorkshire, was raised to a ceiling of about 20 miles to avoid damage to aircraft electronics.

Our dangerous wired world

Wired-up homes in the Western world, connected to the mains, are vulnerable whenever thunderstorms rage. This is why special power surge protection devices are recommended for computers. Computers themselves can produce radio waves, and manufacturers try to ensure that they do not malfunction because of the electronic soup of emissions they operate in. Any burst of electronic 'smog' can have a varying and unpredictable effect, depending on circumstances and the time of day. Computers and attached printers can 'crash' for no apparent reason because the connecting wires might be picking up electric signals from overhead fluorescent lighting. A man in Vancouver was struck by lightning while listening to his iPod; the lightning hit a tree and hurled the thirty-seven-year-old about 8 feet. He suffered burns, perforated eardrums and a broken jaw when the lightning tracked along to the earplugs. Similar symptoms were experienced by a woman in Alabama, when lightning passed into the handset through an outdoor phone cable while she was making a phone call. Said Nathalie Matthews: 'I felt a pain in my jaw, through my ears and down my spine.' A man in the US state of Virginia experienced an identical fate with his phone, suffering burns to his face and arm.[32]

Typical electrical and appliance damage from lightning strikes

1960   In Missouri a bolt had triggered a power surge in one family's home that made the electric clock begin to run backwards. The wiring of the house became completely fused and relinked, so that light switches no longer operated the lights they were supposed to.[33]

1995   In July, several phone operators at a Kirkcaldy, Scotland, telephone exchange were rushed to hospital after the building was hit by lightning.[34]

2005   The US National Weather Service experts in Phoenix,

Arizona, investigated a powerful lightning strike that 'sounded like dynamite exploding', damaging twenty-four homes in one suburb in the summer. The arcing strike produced intense heat which exploded underground wires, including television cables, while a fireball erupted through the soil and spewed dirt and ash around. It damaged homes, trees and parked vehicles.[35]

2006   In Tennessee in June, Carl Schrader was sheltering from a thunderstorm indoors when lightning struck outdoor cables and blew out the entire electrical circuit of the house.[36]

2007   During a storm in West Lothian on 1 July, a bolt hit a street with such a deafening crack of thunder that it shook nearby houses, producing a ball of 'glowing energy', which appeared over a computer in one family's living room.[37]

2007   In July, a television set exploded in a house in Grimsby when lightning struck the aerial on the roof. It shot down the cable into the set, caused a gas meter to explode, which in turn set the entire ground floor of the house on fire.[38]

Bizarre lightning tragedies

According to Dr Elisabeth Gourbière of Electricie de France, only 20 per cent die immediately as the result of a lightning strike, but people out in the open remain vulnerable. Common injuries include shattered bones and burst eardrums. People can get bowled over, and can get their limbs dislocated as a direct result of muscle contractions. A US schoolgirl was killed in 1974 when a lightning strike vaporized a tree she was walking past, and the exploding debris fractured her skull.[39] In the summer of 2009 a man was sheltering beneath a tree in Northamptonshire when a lightning bolt travelled through his body. It burst both his eardrums and he erupted in flames, suffering 11 per cent burns. He knew only what had happened to him when he woke up the next day in intensive care.[40]

Most of the electrical current from the strike 'splashes' over the body rather than passing straight through it. Skin burns are the most likely injury that people suffer from such strikes. The top layer of the skin, especially when it has become moist through sweat or rain, is, unfortunately, a good conductor of electricity. Tooth fillings have been known to melt, and if you have some metal touching your body you are vulnerable, said Paul Taylor, from the UK's Met Office in Exeter. Dr Eric Hefferman at Vancouver General Hospital has said that the combination of sweat and metal earphones directing the current through the patient's head can be harmful.[41] People get burns from keys, jewellery, watches, coins and mobile phones if any of them have been struck by lightning, says Giles Harrison from the Met Dept at the University of Reading.

Robert Davidson was struck in 1979 after stopping by the road on his motorcycle. He told how a new battery-driven watch he owned burnt out in days, presumably as the result of some altered chemico-electrical circuit in his body, whereas his old battered manual watch that he had worn while struck continued to function normally. In some cases, a victim's moist clothes have been known to 'explode' off the body, and this seems particularly to apply to shoes because of the way the bolt travels downwards through the body and is earthed by the feet. For example, in 1969 a man in Missouri, aged thirty-one, emerged from his truck to walk up his driveway in Lawson and was struck so hard by lightning that he was knocked unconscious, only to find that his boots had been blown off. He had been thrown several yards into his neighbour's wall, and the cash in his pockets had merged into a single coin sculpture. He remembers when the bolt struck that he was suffused in a flash of extremely bright white light.[42] In another alarming case, a man also had his shoes blown off in May 2000, as he was playing golf in Cape Cod, Florida. He was in a coma for four weeks, and needed months of rehabilitation before he could eventually walk again unaided.[43]

Further, some 70 per cent of victims sustain either neurological or physical damage.[44] The area that controls emotion and personality is most often affected. This is probably because the electricity travels up the network of nerves and arteries leading to the brain. Debilitating fatigue, momentary loss of memory and difficulty in concentrating often occur. The circuitry of the brain that controls breathing may also be affected.

Some people develop an extreme sensitivity to alternating currents, and the discomfort is so intense that they have to live without the convenience of electrical appliances for some weeks afterwards. One girl was struck by lightning while on her phone and suffered cardiac arrest, and was left with severe physical difficulties and brain damage. According to spokesmen from the Northwick Park Hospital in London, the cellphone made her injuries worse by directing the lightning current inside her body.

Others who have been knocked out and burned severely by lightning speak of alien experiences. Many victims feel as if they have lost time, and find themselves recovering from the blackout a small distance from where they were. They seemed to have no awareness of what had happened. Dave Kezerle, in August 1977 while walking through woodland in Illinois, said he had a fisheye vision of being in a cylinder that was glowing with a 'beautiful aquamarine'. Then he felt he was being sucked up, and 'little sparks began to pop around me'. Another nearby witness who had to be hospitalized told how the intense flash had enveloped him 'like a fluorescent light bulb, shimmering all over'. [45]

Louis Proud points out that about seventy people die each year in the USA from lightning strikes. It is reckoned that over a person's lifetime, the chances of being struck by lightning just once is one in 750,000. So to be hit seven times in one lifetime would be extraordinary bad luck, with the odds against this happening being 'astronomical', according to Dr Proud. He relates the story of a park ranger who had worked most of his life in the Shenandoah National Park, Virginia, and who was indeed struck seven times between 1942 and 1977. He suffered enormous psychological harm and neurological injuries, with severe burns down one leg and the loss of a toe. In virtually every case either vehicles or metal instruments or technology was near the ranger at the time, from which the lightning bolts ricocheted. In 1969 lightning bounced off trees into the open window of his truck, knocking him unconscious and burning off most of his hair, with his hair catching fire more than once in all these events. [46] A year later he was bowled over after the bolt bounced off a nearby power transformer. After the fourth strike in 1972 he began to feel he was jinxed. He took to crouching down in the back seat of his truck whenever he got caught in a storm. Again, another strike in August 1973 hit him as he got out of his vehicle, and yet again in June 1976. He took his own life in 1983 at the age of seventy-one.

5 Our Lit-up Skies

Akhlesh Lakhtakia, who is an electrical engineer based at Pennsylvania State University, suggests we should be cautious in interpreting what we see in the atmosphere. He published a series of academic papers showing that some astronomical observations could not be entirely reliable because of night sky optical illusions. The varied colours that people, including astronomers, see are merely aspects of sunlight as the sun goes through its phases. This can often mislead, as the sun's magnetic field interacts with Earth's field to create peculiar illusions. Spectacular twilights and a mysterious electric ice-blue glow occurred in many parts of Britain and Northern Ireland in the spring of 2008.

Some people have said they have seen ripples of glowing silver-blue clouds on top of orange-red glows. These unusual 'clouds' are in fact about 50 miles high, but succeed in catching the rays from the sun long after sundown. A dry atmosphere can create spectacular sunsets of vivid orange and an afterglow.

Light refracts automatically when it passes into other materials such as water or glass, but light can bend so far that its spectrum of colours is shifted. The brightest star is Sirius, and when it is low on the horizon its light can fluctuate between green, red and orange as it passes through the maximum depths of our atmosphere.

Rainbows are caused when sunlight hits a raindrop, and does not move as fast through the water as it does through the atmosphere. The sunbeams twist a little on the way into the rain cloud, and twist again as they leave the raindrop to return to the air. So when light hits the rain at just the right twisting angle, it is refracted through a raindrop and into the colour spectrum.

But you can get rainbows without a drop of rain. A curved

coloured sliver of cloud appeared over Copthorne, near Crawley, in Sussex, in August 2009, looking like an upside down rainbow. Another was seen 'hanging upside down' near Edinburgh on 27 May 2009. These were not actual rainbows but arose from the strange light deflections caused by the cirrus clouds that coasted some 20,000 feet above,[1] and are known as *circumhorizon arcs*.

An unusual borealis deflection of the northern lights becomes a *circumzenithal arc*. This arc is sometimes known as a 'fire rainbow'. Such was seen in the USA on the Washington–Idaho border in early June 2006. Dr Jonathan Fox, of the US National Weather Service, in Spokane, in Washington, said it was 'even more spectacular than the Northern Lights ... it forms in only very rare situations. This is the first one I've ever seen. It was a breathtaking sight and it hung around for about an hour.' [2]

The *Hexham Courant* published a photo of a faint 'rainbow' taken over Hexham, in Northumberland, at the end of January 2002. This photo received wide publicity, with press pictures showing the phenomenon stretching 42 degrees across the night sky. A witness said she thought the arc of light was a laser beam, but it was, in fact a 'moonbow', created by a full moon shining through a shower of rain.

A sun *pillar* is a vertical column of light that occurs when a ray of sunlight beams upwards behind a cumulus cloud, or through a veil of thin cirrus clouds which are so high that their water is frozen into millions of ice crystals. A red shaft of light once rose up from the Sierra Nevada mountains of California, which was illuminated by the sun setting behind the mountains, and was shown as a 'Picture of the Day' on a website (http://antwerp.gsfc.nsas.gov/apod/ap060205.html).

When the sun tries to shine through a veil of hazy cirrus clouds and the remains of aircraft contrails, or if the sunlight is bent at 22 degrees, it can create an illusion known as a 'sundog'. This is an optical phenomenon that appears as a rainbow-coloured BOL, often with a 'tail' trailing behind it, and always streaking away from the sun, invariably dead straight, short and stubby.

There is a confusion, however, about 'sundogs'. A ball of orange and yellow light was seen to be 'moving fast', according to Matthew Pinless, aged thirty-one, who saw an 'object' at Cheltenham, Gloucestershire, in late August 2009. This was probably not a sundog, but was described as such in the media. It was illustrated in the *Daily Mail*.[3] Neither could the perfect circular rainbow that was photographed in the skies over Malaysia on 6 August 2007 have been, as suggested, a sundog. The same goes for a fireball seen over Leeds and filmed on a mobile phone, with clear pictures shown in the *UFO Data* magazine of March–April 2007.[4] Again, an astronomer from New Chapel Observatory in Yorkshire wrongly described it as a 'sundog', yet clearly the camera-phone showed an unlikely curving, smoky tail.

Exotic clouds

Exotic visions frequently occur in our skies. *Noctilucent clouds* (NLCs), also called night-shining clouds (NLCs), are some of the most beautiful and dramatic sights on Earth. They are best seen at dawn or twilight when they catch the sun's reflective rays, especially the *angle* of the sun's rays, which makes them appear as shimmering silvery-blue ribbons. In the mesosphere – about 30 to 50 miles above the surface – temperatures drop to -200 degrees F, with 200,000 times less atmosphere than we breathe. Hence NLCs remain in sunlight while the other clouds in the troposphere – the lowest portion from 1 mile to 30 miles up – are in darkness in Earth's shadow. In 1885, NLCs, shimmering in metallic blue colours, seemed to be appearing after the huge eruption of Krakatoa, about 50 miles high in the air, hovering on the edge of space in the mesophere. The volcanic dust veil also interfered with global sunlight patterns to create spectacularly coloured iridescent skies around the world.

Noctilucent clouds may also be linked to the sun's rotation, according to a recent theory. Charles Robert of the University of Bremen and his colleagues found that the clouds appear to wax and wane over a twenty-seven-day cycle, the same time the sun takes to rotate on its axis.[5] NLCs were studied in 2007 by Nasa's Aeronomy of the Ice in the Mesophere (AIM) spacecraft after it was noticed that the world's highest noctilucents have been appearing in unusu-ally large numbers in the 2000s. Rare mother-of-pearl coloured clouds known as *nacreous*, or polar stratospheric clouds, are produced by air rising over Arctic and Antarctic mountains during winter, and form in low temperatures of -112 degrees F at 30 miles

above ground. A sudden drop in temperature could trigger disturbances in the upper atmosphere, precipitating ice crystals that bend
the sunlight into darkened bands, and sunlight enters the near-
vertical face of six-sided (hexagonal) crystals to create the illusion.

But there is an additional effect with atmospheric crystals. Walt
Petersen and his team at the University of Alabama in Huntsville
analysed data from Nasa's TRIMM satellite, which monitors both
lightning flashes and the amount of nacreous ice crystals – often
known as 'ice halos' – in clouds. Petersen and his team found that
myriads of tiny pieces of ice inside high-altitude thundercloud tend
to bang into one another, and this sparks off static electricity in the
polar clouds, so much so that spectacular shows of red, orange,
blues and greens can sometimes be seen veering hundreds of square
miles across the sky.[6]

The *altocumulus lenticularis* cloud is 'lenticular' shaped, i.e. the
shape of lentils, or discs. Some disc-shaped clouds inevitably
become confused with UFOs. In the Massif Centrale in France
during the 1950s UFOs were often said to be cloud 'cigars'. The
cloud condenses on the crest of these moisture-laden waves but
disappears as the wave dips down and dries out, leaving behind a
smooth, disc-shaped cloud that looks stationary, and even seems to
hover. It grows more spooky if the sun shines from behind it.
Similarly, a sunset in early October 2004 after a rainstorm in
Britain picked out broken patches of altocumulus and cumulus
clouds which showed up orange and pink. Parts of the clouds were
arranged in a fishscale pattern. This was the phenomenon known
as *iridescence*. The light, instead of refracting, was *diffracted*. It is
bent slightly as it passes around the edge of the cloud droplets, with
the light waves interacting to create coloured patterns. Iridescence
usually has a metallic quality, rather like oily residues that can be
seen on water surfaces. Iridescence is named after Iris, the Greek
goddess who passed messages between mortals and the gods. As
she ran back and forth, her multicoloured robes shimmered.

There are other reports of strange aerial colours, probably caused
by the scattering of the colour spectrum away from the blue. A sky
that turns green is often a precursor of tornadoes, and green clouds
were seen in swarms of 148 tornadoes in the USA on the 3 and 4
April 1974. Cumulus clouds cast a remarkable green light on 20
August 2009 over Dumfries, in the south-west of Scotland.

A peculiar cloud formation was spotted off the coast of Hokkaido by a Japanese coastguard aircraft. It looked like long tubular lines of carpet material lying parallel to one or other. These may have been 'roll' clouds, made by waves of 'gravity trains', which behave like waves from a boat. They are more often seen by pilots, especially, for some reason, in Australia, and are known as the 'Morning Glory'. They often drift in near dawn, and can be 620 miles long.[7]

Cylindrical clouds, similar to 'roll' clouds, are fashioned by fast air currents, when air rising through a region of high humidity abruptly descends once more, and water condenses out to form droplets, and the vision becomes elongated because of the movement of air currents. They are usually spotted on the leeward side of mountains, like those in New Guinea or Germany's Rhine Valley. But they are also seen in flat countryside, and are probably the creation of convection currents generated by *heat islands* – localized urban areas that generate more than the overall level of heat.

Aircraft can produce some strange images in the sky. Sometimes 'shock collar' scenarios occur when energetic fields ionize the surrounding air. Cone-shaped white clouds no larger than 100 feet across can be formed as the 'shock collar', or vapour cone, of jet aircraft that are created in certain atmospheric conditions. This in turn produces the strange variations of coloured light sometimes seen around aircraft fuselages.[8] A dark leading trail can sometimes be seen *in front* of an aircraft in misty weather. They are thought to be formed by a shadow of the jet trail behind the aircraft, caused by the sun beaming on to the misty atmosphere through which the aircraft is about to pass.[9] A photograph of a similar sight appeared in the *Daily Mail* on 28 May 2009. It was captured on camera at the Jones Beach Air Show in Wantagh, New York, as a US Navy Super Hornet seemed to be passing through it. This time it was caused by pressure differences around the aircraft, and the shape of the Super Cornet itself, which can travel at nearly 1,400 mph, which encourages the cone to form.

The riddle of the 'cloud-UFOs'

Some apparitions in the sky, as we have seen with the *lenticularis*, could possibly be UAPs that have crossed over the line into the UFO category, and many of them have been described in mainstream science journals. Hazy UFOs can cause confusion by becoming part of the overlapping UFO/UAP/sundog syndrome, looking like fat blobs of light, often seen with a blurred, colour-rinsed solid edging. A village of 800 on the moors above Matlock, in Derbyshire, has a long tradition of seeing unconventional UFOs. In the years 1999 to 2001, there were separate sightings from the villages and moors of pink BOLs. A slug-like glowing object, which turned black and flew off, was reported over Liverpool in June 2006.[10] I myself saw such a phenomenon which I thought was the product of aircraft exhaust. It was in the early evening of 30 October 2007, as the sun was beginning to set over south-east London. It was isolated in a clear sky, and was illuminated with an orange hue, bulbous and slug-like. It took about twenty-five minutes to slowly sink to the horizon, but while doing so at one stage it became momentarily stretched and faster moving. This ruled out either a descending meteorite, an aircraft contrail or a sundog.

Some UFOs also look like clouds because of the low level of light coming from them. Sometimes a brightly illuminated UFO emerges from the clouds, and is surrounded by 'luminous vapour'.[11] References have been made to 'cloud cigars' by well-known ufologists Aimé Michel and Jacques Vallée, both prominent figures in the 1960s and 1970s. Several were seen over France in 1952 and 1954. Many strange UAPs, possibly 'cloud UFOs', have been seen in the so-called 'Pennine Window', an upland area roughly defined by drawing a triangle between the cities of Manchester, Sheffield and Leeds, a zone that covers much of the northern part of the Peak District National Park. The Peaks lie below a major international flight path that is an occasional training ground for low-flying military planes. High on the moors one can easily fall victim to illusions, and low mists can distort images, and there are plausible, naturalistic, explanations for many of them.

A Mexican researcher, Jaime Maussan, has commented on videos showing Mexican air force pilots using infra-red technology that seemed to reveal UFOs, and some footage of 'snake-like' objects high in the sky were taken.[12] A worm-like illumination was shown in UFO Data magazine for March 2007.

But many UAPs continue to baffle as they hover, literally, between natural and unnatural categories. On 4 October 1996 at

Glyn Ceiriog, Wales, grey-white light flickering above clouds was seen, a phenomenon that accelerated in a zigzag fashion. The light was about 8 feet in diameter, with a smaller light at the centre and another light around the edge, which was rotating. There were lines of light travelling in a circular motion, looking like the underneath of a mushroom.[13] A man in Derby saw a bright white object late on the evening of 16 August 1987. It was moving in a downward arc, and left bright blobs behind. He doubted it was a meteorite. A woman on the same night saw a 'massive' white light with 'brilliant lights' on either side of it, and a red glow underneath. It looked like an illuminated double-decker bus. Others in Derby saw a pie-shaped UFO, and others saw 'circular lights'.[14]

A bizarre incident involving wind turbine damage at Conisholme, in Lincolnshire, was widely covered in the British media in January 2009.[15] Some saw orangey-yellow spheres skimming across the sky in the small hours. Another report spoke of a 'massive ball of light' with 'tentacles going right down to the ground'. Then, at 4 a.m., there was an ear-splitting bang. One of the turbines had one of its 60-foot blades ripped off, which was never found, and another was knocked into a twisted shape and rendered useless. John Harrison, of nearby Saltfleetby, saw the BOL and its 'tentacles' over the farm. The inevitable 'falling block of ice theory' from a plane soon emerged, but windfarm builder Ecotricity ruled this out. Alternatively, it could have been lightning, but again Ecotricity ruled this out, as they did a helicopter strike.[16]

Indeed, many UFOs shown on the internet seem to be either vastly exaggerated sundogs, or auroras, or bizarre and unrealistic-looking clouds with massive dimensions. A huge and menacing grey disc, perfectly circular, partly cloud and partly solid, hung motionless in Latin American skies, although the exact location was not identified. It had strange wisps of other gases or clouds drifting from its underside. In another video clip a deep orange-red and very brightly illuminated cloud-spiral was seen over Canadian skies at dusk, but the source of the light was not clear. The faintly pink clouds in the distance, reflecting light from the fading twilight, could be used as a comparison to show that the light did not come from the sun.

Several photos of 'ball plasma', with reddish sundog outlines (with some appearing to be double-jointed) were taken over the Vale of Pewsey, in Wiltshire in 2001 and during the summer of 2003, and shown in *UFO Magazine*. In fact, hundreds were seen during the 1990s and early 2000s in that region.[17] In the transition stage from a solid to a gaseous or translucent stage, some UAPs seem – judging from eyewitness reports – to become a haze or a 'mist'. A UAP was filmed over a theme park in the US state of Virginia in July 2009. It was shown on the *Daily Telegraph* online video site on 3 July 2009, and on YouTube. The video shows a perfectly circular ring shrouded in a haze, but apparently hovering and rotating, about 200 feet across. The woman who filmed the video said the object, although appearing as hazy or smoky in the moving images, was clearer in reality, and the sharpness of the outline makes natural phenomena unlikely. Cameron Pack, a UFO investigator, said such rings replicated what had been seen in Orlando and Florida in 1999, and at Fort Belvoir in the 1950s. The haze could be the outer part of the UFO in between a solid and dissolving phase.

Other internet video clips show a giant circle of a 'light spectre' that was filmed over Moscow at sunset. Seldom do clouds have such an obvious holed-out doughnut effect. In Indonesia in February 2010 a massive illuminated cloud was filmed with what looked like an orange disc perched on top of an apparently genuine cloud, but it was unclear whether a giant UFO was being viewed or not. It is extremely unusual for clouds to be two-toned like this one was.

On 10 September 2006, over north-west Santa Monica, a large, self-luminous bright sphere above three or four smaller and less illuminated ones, stacked on top of one another, were seen. Another looked as if it contained large spheres that were linked up or covered with something, like a magnet or ion plasma sheath, 1,000 feet in the air. The object was at least 100 feet long. This event was videotaped and shown in *UFO Data* magazine. It was described as slowly gliding, before stopping and reversing its direction. Its appearance then began to change shape from a long curving tube with a bright sphere at each end to one large sphere below, from which began to emerge small white spheres.[18]

These multi-hybrid UFOs – partly atmospheric phenomena, partly paranormal, and partly 'flying saucer' – indicate not only the ability to 'morph' from solid to less solid. An RAF man in October 1952, on duty on an overcast day, saw a 'peculiar black cloud' from which a sphere-shaped object emerged.[19] In November 2006 a

UFO incident occurred at O'Hare airport in Chicago. A large circular and metallic object appeared to be hovering and spinning. However, one aircraft mechanic at the airport described it not as metallic, nor as precisely circular, but as a 'dark grey, hazy, round object'. This could possibly be atmospheric phenomena, because the mechanic added that the object seemed to be 'trying to stay close to the cloud cover', which was low that day. Another witness said the UAP was at least 700 feet off the ground and below the cloud base. The top of the object was well defined, but the edges were blurred and distorted by a heat haze.[20]

Leading examples of reported cloud-UFOs

1853 The *Illustrated London News* reported a bright beam of translucent light in the Huddersfield area 'springing from above a cloud'.[21]

1853 In September at Retford, Yorkshire, people saw a 'nebulous band emitting a bright aurora light and slowly moving from east to west'. The stars apparently shone through this vision: it became stationary and assumed the configuration of an S, and was assumed to be the tail of a comet. The nucleus was 'brighter than the moon, but was seen in daylight, and was nearly globular', largely red, with a blue and yellow tint round the outside,[22] although this doesn't fit in with any cloud or defraction effect.

1931 The crew of a steamship crossing the North Atlantic on 24 May at night suddenly saw the sea and sky brilliantly lit for about three seconds 'with a flickering purplish light, which did not appear to emanate from any particular point'.[23]

1932 Over Cache Lake, Ontario, in July 'a very long, low narrow, tenuous cloud, resembling a squall cloud', was seen. It was 200 feet long, and emitted a mysterious rumbling noise, although there were no lightning flashes.[24]

1936 In Tauranga, New Zealand, in the early summer, dark clouds suddenly began to glow with a bright greenish light at the top, which lit up the rest of the cloud bank. A jagged stroke of lightning then flattened itself out horizontally, and zapped into the green light.[25]

1949 A witness in Yellowstone National Park in early September watched a 'hazy patch of blue light', about 50 yards by 250 yards, sweep across a marsh. It was moving fast under a low cloud, and when it got close to the ground it seemed to 'flow around' everything that it came into contact with, 'coating it with a strange pulsating light'. It covered his car and his person, resulting in a tingling sensation in his scalp.[26]

1959 On 1 December, a bright blue light was seen by Capt J. Williams on the MV *Grevean*, sailing through stiff winds under low cloud in the mid-Atlantic. He described it as 'bright blue diffused light which suddenly grew in size ...'[27]

1964 In Ardmore, Oklahoma, on 9 April, just after 8 p.m., a large object was seen in the sky which had 'windows', and seemed to dissolve in mid air after a few seconds. Two days later another UFO, looking gaseous or cloud-like, came into view over New York State. This time a long dark object remained in view for forty-five minutes, but it was in a strangely inverted position, with wisps of smoke streaming out of the main body of a black cloud.[28]

1971 In June a 'green misty thing' was seen over Warminster, in Wiltshire, just above ground level at night. Several witnesses said it kept shifting its ground.[29]

Early At Thundersley, Essex, witnesses saw a slow-moving 'white
1970s stream of light' coming down from the clouds from a 'sort of grey blob'.[30]

1974 On 14 August , near the town of Aveley in Essex, in misty weather, a family saw from their car a bright star that followed them before they encountered a large 'greenish' fog or mist as they rounded a bend. As the car entered this mist, the car shuddered violently, and its systems failed.[31]

1990 In the West Country, in July, two witnesses saw a swirling white shape, rotating in a clockwise direction.[32]

1991 Rolling luminous mist effects were seen at ground level during bouts of rain.[33]

1994 Jean-Charles Duboe, a retired air force captain, reported that on 28 January, during an Air France flight 3532 near Paris, a 'huge flying disc' flying at 35,000 feet at a distance of about 25 miles away could be seen. It was 'red-brown with blurred outlines', then 'became transparent' and soon disappeared.[34]

2001 Camcorder footage taken in October at Slaley, near Bonsall, revealed an 'eerie fog' that appeared, plus a circular cloud composed of 'rainbow lights'. Other witnesses said it looked like a giant disc with illuminated coloured rings rotating inside the disc, with a 'bite taken out of the bottom' (the 'notch' syndrome – (see Chapter 14).[35]

1998 On 21 April at night in Liverpool, a UFO about 200 feet off the ground with a 'sort of mist around the edge' was seen; it was making a humming noise. A dim light also rotated around the edge every ten seconds.[36]

2002 A Washington photographer captured UFO images on Capitol Hill in July, both on photo and video. He recorded green glowing objects, with some leaving a blue signature.[37]

2005 On 3 October, near Scammonden Dam, a collection of about fourteen UFOs were seen as faint grey oval shapes, sweeping across Holme Valley before rising and forming a perfect circle before dispersing at high speed.[38]

2006 In April, a couple saw what looked like a narrow band of shimmering mist moving in an undulating manner over trees and houses, gaining in brightness.[39]

2006 On 28 June at Bigbury, in south Devon, near Dartmoor, an orange ball hovering over a corn field was seen, before shooting off. Another softer-looking object had four lights inside it, glowing a dull greyish-white, fading in and out.[40]

2006 On 1 July , near Gisburn, in Lancashire, an orange BOL was seen, stopped, then joined another, until seven appeared.[41] Fifteen days later at Chorlton, South Manchester, three silver orbs were seen at a high altitude, on a hot day with clear skies, lining up and then moving in synchrony into a 90-degree turn. A plane then flew below, leaving a vapour trail, which itself attracted a metallic object, smaller than the plane.[42]

2007 In April, over Dublin, an object was seen situated in between three lights. Overall the object had no definite outline, but appeared as a haze of grey – like a shadow or dark cloud.[43]

2010 On 1 January, over Abu Dhabi, a giant blue cloud in the shape of a perfect circle was videoed, and shown on the internet.

2010 In Chile, a large crimson illuminated ring could be seen on an internet clip; it was sinking over the horizon at sunset on the night of 26 February, but it seemed to be separated from the background.

Dusty visions

At this stage we could ask why Earth's sky is blue, and why does it occasionally change colour? Unfortunately, before we give a reasoned answer to this we should be aware that we are not always told the truth about the colour of solar system skies, or indeed about other aspects of space science. Very few scientists or science writers, for example, would include UFO-type examples of strange atmospheric phenomena precisely because it would be regarded as 'anomalous' (the exception to the rule is the weather journalist at *The Times*, Paul Simons, who has been reporting on such aerial phenomena for years, many of whose examples I have incorporated in these pages).

We often do not know whether the science is wrong or the 'politics' (in other words, the way that science is mediated before it reaches the public) is wrong. 'Rayleigh scattering' is the defraction of light by particles of *any* gas that are smaller than the wavelengths of light, and has to do with the amount of dust in the atmosphere, which will tend to have a hazing, or screening effect, rather than a tinting effect. Different colour *frequencies* will scatter by different amounts, depending on the gaseous nature of the sky itself. It is a function of the electric polarization of particles. The reason the sky is blue is that the wavelength of blue is short, so it scatters more. The reddening of sunlight is intensified near the horizon because the volume of air is high. More importantly, when sunlight hits the atoms of air high up in the atmosphere, some blue light is scattered out, and the remaining light, robbed of its blueness, will appear rosier than otherwise. Basically the sky is violet, but our eyes mainly see blue during bright sunlit days. The sun appears to have a yellow colour, although out in space its nuclear reactions would appear to be white or blue-white, and the 'sky' would appear black.

However, on the internet you can see diametrically opposed views about the Martian sky. According to some, red dust makes the

Martian sky rusty-coloured, because of the iron oxide on the Martian surface that is blown by fast winds into the sky. But many scientists believe that the Martian sky is blue, like Earth's is, and the first colour pictures of Mars beamed back in the 1970s showed this, but the colour was stealthily (i.e. without explanation) adjusted afterwards. (The blogging site Etemenanki.goroadachi.com explains the motives and technicalities of this in some detail.)

But this is not to argue that different hues seen in Earth's skies are not due to pollution. A considerable amount of the pollution that alters the hues in the atmosphere surprisingly comes from space. Dust is in a band close to the plane of the Earth's orbit around the sun. The effect of both space dust and air pollution can dramatically alter the visual understanding of both astronomers and people on the ground who think that they have seen UFOs. The cone of misty light that can sometimes be seen as the sun rises in the east, and creates an ethereal glow, is actually the *zodiacal* light. It is seen all year round, but is much brighter in the tropics. It is created by sunlight scattered off clouds of cosmic dust, or from tiny bits of comets and asteroids.

Similarly, large coloured rings (sometimes known as 'halos') appear around the moon when it reaches its perigee, and can be seen through thin clouds. Tiny droplets of water in the clouds succeed in scattering moonlight into what appears to be rings of coloured light. A soft smudge of light opposite to the position of the sun, as the space dust scatters sunlight, produces startling effects similar to the sprite auroras referred to earlier. It lasts no more than a second or so. The sun can occasionally appear blue during forest fires, as tiny soot particles of the same specific size filter out the orange of the sun's rays in the same way that oxygen and nitrogen molecules do. This happened in the Californian wildfires in June 2008 and in Scotland in September 1950, as a result of a pollution haze.

Weird light scattering is also due to seismic eruptions. J. M. W. Turner's sunsets in his paintings were caused by volcanic ash and dust that spread around the globe, as the dust scatters sunlight when the sun is low on the horizon. Volcanic light scattering is more common over deserts, but it can also take place in higher latitudes. Scientists can even tell how much volcanic dust there was in historical skies by examining the Old Masters, painted between the years 1500 and 1900, such as those by Turner, Rubens, Gainsborough, Rembrandt and Hogarth. Turner recorded the skies after three major eruptions, one of which was Tambora in 1815.

Vapour erupting from a volcano can also turn to ice at higher latitudes, and fragments of rock and ash particles shoot upwards in the towering plume, with the sparks ejected from the vent glittering brightly. In Chile an eruption met up disastrously with a lightning storm in May 2008. Tons of dust and ash from the eruption of Chaiten's ashes poured into the night sky. Stunning sunsets also appeared after Kasatochi in Alaska blew up a few months later. It scattered yellow light particles to the horizon, which then blended into orange, red and purple.[44] And there were 'fantastic' sunsets as Mount Redoubt erupted in south Alaska on 22 March 2009. Satellites revealed the plume sweeping over North America and the North Atlantic. Illuminated ash drifted down to Anchorage, and aircraft were re-routed.

Natural atmospheric glows

Space affects the path of light because mass and energy distort spacetime, creating warps and dents. When the light from stars travels through our atmosphere, it bends, rather in the way that putting a stick in a pond appears to make it bend from the surface of the water downwards. This is also the basis of gravitational lensing, with some massive object in the way that prevents light from reaching Earth, and thus into the range of human vision. An amateur photographer, Jan Lameer, caught the sun rising over the top of Monte Viso in the Alps, when it appeared momentarily to turn blue. This could only happen in an extremely pristine and calm air environment, when the layers of air settle and bend the rays of sunlight like a giant lens. The spectrum is teased out to produce a flash of blue lasting only a few seconds.

The ionosphere, the geomagnetic field and the 'magnetotail' – the long trail of bright illumination coming from the field itself – are all involved in spectacular displays and these can so often result in confusing UAP reports. Earth-bound gases, whether in the open air or trapped within a fluorescent tube, can produce light as a result of external stimuli, such as when a spark energizes them.

Atmospheric nitrogen under certain conditions can exaggerate aurora phenomena. When this happens it can produce a soft white glow that can last some minutes after a blob of electrified particles manifests itself.

A burst of green light at the setting sun is sometimes produced through refraction of the atmosphere. A faint green glow is called the *gegenschein*, German for 'counter-glow'. A green light flash from the setting sun was also described in the 1882 Jules Verne novel *The Green Ray*. The green was followed by blue, but tended to get scattered in the atmosphere. An intense beam of blue light also came from the top of the sun as it set near Glenrothes in Fife, Scotland, on 4 January 2006, and was captured on film. Lord Kelvin in 1899 saw a blue light against the sky on the southern face of Mont Blanc when he was staying at a hotel on the French Alps in the summer. He said that the flash lasted 'only one-twentieth of a second' before it became 'dazzlingly white ...'

A lunar eclipse – with the Earth passing between the moon and the sun – on 3 March 2007 saw the moon become blood-red. Light that was scattered through the Earth's atmosphere was altered to predominantly red wavelengths, which reflected off the lunar surface.[45] Les Cowley of www.atoptics.co.uk said that the rays would be almost white but for the ozone effect which absorbs some red light. In other words, the moon was not totally blacked out by the eclipse.[46] When a solar eclipse took place in early April 2006 in the Libyan Sahara, the sun seemed ringed by fire in a very dark sky, with Venus and Mercury shining brightly. The horizon was pink, orange and yellow in the twilight.

Solar eclipses also produce their own weather phenomena, and the sudden cooling of the desert air with the eclipse. This causes winds and breezes. But people were puzzled to see mysterious dark stripes beamed high across the sky just twenty minutes before a full eclipse in August 1999, seen over the Channel Islands, which ran parallel with the horizon and crossed the path of the moon. The stripes lay above and below the sun and ran two-thirds of the way across the sky. Some scientists disputed this and suggested instead that the bands were merely the shadows of contrails.

The connection between electricity pylons and light was successfully demonstrated in a British tabloid article in early 2004. A phosphorous coating on the inside of the fluorescent bulb converts the UV light to visible light through a chain reaction involving gases inside the bulb. This explains why, when some 1,300 fluorescent light tubes were planted in symmetrical rows in fields immediately surrounding a pylon complex in south Gloucestershire, they started to glow. The overhead cables were capable of producing 400,000 volts of electricity, and when this power seeped out into the surrounding air it was able to faintly illuminate the ordinary domestic tubes by exciting the inert argon gas inside which reacts with the tube's inner phosphorous coating. Even individuals could get the tubes to glow by waving one of them beneath the pylons, because both holding the tube or planting it in the ground earths the device and completes the circuit.

Significant clues from this little local experiment spurred some scientists to draw conclusions about other illuminated phenomena. Something in the air could make things fluoresce in the dark that were more than 'natural atmospheric glows'. This power could take on a spherical shape from time to time, and may actually be connected with phosphorus or some other gas – even gas that had been broken down into its sub-atomic particles.

There is much erudite speculation, of course, in this shadowy world of energy physics. But can science really explain what 'balls of light' are made of? We will examine this aspect of illumination in the next chapter.

6 The 'Ball of Light' Syndrome

A man tramping across the peat bogs in a tranquil valley in Donegal in Ireland in 1868 saw what he thought was a storm brewing. Pondering whether to return home, he then saw an amazing sight. Floating down the hillside was a glowing red orb. It was about 2 feet wide, and silently bobbed over a stream. It seemed to shrink in size as it landed on the soil, before finally disappearing. There was no sound, no explosion or fire. Yet the ground beneath its path had been churned up violently, and indeed had torn away about 75 feet of the bank of a stream. The hole was still there when alarmed local people visited the site.[1]

So what actually happened in Donegal? Unfortunately, over the years, as similar reports made the rounds, one has to hazard a guess about which of the LITS, BOLs or atmospheric electrical phenomena best fits the bill. Most of the light phenomena I am describing in this chapter are to be seen close to the ground and are to be distinguished from 'sundogs' and 'cloud-UFOs', and the weird range of colours displayed by many BOLs that do not seem to correspond to the heat-related frequencies of bolides. But often BOLs elide confusingly across categories. For example, at Rowarth Peak, in Derbyshire, on the night of 10 September 2001, bright lights were observed across the hillside. They were hovering and 'dancing' just above the ground, then appeared to ascend into the cloud, 'lighting up the whole area around them.[2]

One category of mini lights is *bead lightning*. This is more familiarly associated with BOLs, as a variety of energy coming from lightning bolts. They used to be known as St Martin's Beads, after the parish of St Martin-le-Grand in London. The area was known for shops selling cheap, imitation jewellery. Dermot MacManus, in

The Middle Kingdom, 1959, speaks of 'fairy lights' because they are more like Christmas tree lights. A string of them was seen to fly from a 'fairy fort' near Trillium near Castlebar in the west of Ireland.[3]

A 'white globe', associated with a nearby thunderstorm, struck the town of Brecon, in South Wales, on 25 April 2004. In one house it 'blew away the chimney and showered bricks over 100 feet away'. The BOL broke into a shower of small lightning bolts, and travelled a quarter of a mile across the roofs of other buildings, before exploding loudly. Another BOL quickly followed suit, 'shooting through an office with a loud crackling noise'.[4]

As they are frequently seen during thunderstorms, many bead lightning effects are connected with stormy or moist weather, or weather that is heavily charged with electrical static. In 1935 in Johannesburg bright luminous beads were clearly visible after a lightning flash had faded away. A witness saw 'an almost dazzlingly bright', fattish lightning bolt over Weston-Super-Mare in May 2006, and reported it to the Tornado and Storm Research Organisation (TORRO), based in Warrington, Cheshire. He said that the 'bolt' broke up into a chain of dozens of separate beads, following the line of the lightning strike. Varieties of bead lightning sightings might also account for the hundreds of lights heading out to sea along the Kent coast and captured on video, with some detaching themselves from the main group, dropping back, and then quickly catching up again.[5]

St Elmo's, similar to bead lightning, was often seen in the days of sailing ships during stormy weather, with lights playing around the masts and rigging, but the explanation often varies. Herman Melville, in his novel *Moby Dick*, tells the story of a wooden whaling ship that meets a typhoon south-east of Japan and is plagued with crashing thunder, lightning and displays of St Elmo's fire. Subsequently the magnetism of the ship's compass needle is discovered to be reversed. Melville suggests this is not uncommon. During a violent storm, 'three balls of fire', according to *The Times*, struck the upper rigging of the HMS *Warren Hasting* which was moored off Portsmouth on 17 February 1809, and hurled a man down, killing him.[6]

These sightings might simply be another interpretation of other visual events, as I have suggested. Stueart Campbell is an architect and a former UFO sceptical researcher, and believes most UFOs are ball lightning,[7] only because he has expanded the size of the globular object seen in the phenomenon beyond what other scientists believe.

So, to return to the first example of BOLs in this chapter, we have to make an educated guess as to the nature of what was seen in Donegal in 1868. We learn that a storm was 'brewing', but we are unclear about whether an electrical storm was soon 'raging'. It is difficult, hence, to assess how any damage could have been caused on the scale described. Heat and chemical reactions wouldn't have done that without producing steam in the dank earth. It is doubtful, too, if it was any sort of electrostatic or electrical charge (existing as a purely bizarre electrical phenomenon) because the damp earth would have earthed any charge in just a microsecond.[8] Nevertheless, this strange episode was related to the Royal Society in that year by a local scientist with some knowledge of the area and the people.

The historian Flavius Josephus in AD 80 in Jerusalem mentions the sky darkening, turning to dark green, and a ball of pale blue light emerged from a cloud to speed to Earth.[9] Roman soldiers watched it as it jumped off the roof and fatally struck one of the soldiers with a loud boom.[10] Ball lightning was seen after a hurricane on 2 September 1786, which lasted some forty minutes.[11]

Some scientists say the hues of ball lightning objects go from blinding white to sultry red, with a variety of sizes. Witnesses speak of coloured 'air bubbles'. One was reported in Remenham, Berkshire, in the winter of 1861. And at Ringstead Bay on a particularly sultry day in August 1876, numerous 'billiard-ball' types of lights were seen just above the ground, moving up and down. They were described also as being like 'glowing soap bubbles'. They seemed to grow in number; some reports referred to 'thousands' of them at one time. This apparition lasted about an hour.[12]

The general belief that BOLs are connected with ordinary lightning is reinforced by the number of meteorological agencies that have done research into the phenomenon, or have commented on it. Again, there is the problem of categorization. Met Office, police or official descriptions of fireballs could be what local people refer to as UFOs, because many do not fit the usual description of BOLs. During a thunderstorm in Clarens, Switzerland, in the summer of 1880, a small girl and some adults some distance away in a cemetery, saw something covered in 'sheets of fire', or perhaps 'luminous clouds'. Some described it as an 'electric fluid'. Another fiery object bounced from point to point on an iron railing. When the witnesses approached the apparition they felt a stinging sensation on their faces, as if 'gravel' had been thrown at them, and static passed from the fingertips of each member of the group.[13]

In the scientific section of the USAF's 1960s *Project Blue Book* report on UFOs, the lightning analogy was confirmed as the default position, seemingly to occur following lightning strikes: 'At the ground end, the terminal flash is intense, and vapours, smoke or molten material from objects fused at points struck may enhance and extend the duration of incandescence.' Possibly there was an after-image in the human eye. 'Brush discharge' could occur, with 'corona discharges' from pollution particles. Lightning 'streamers' could jump gaps between pipes in houses, causing bright flashes. 'Ball lightning' could arise when lightning bolts go horizontal, said the *Project Blue Book*, explaining how this could initiate 'electrical discharges' and 'luminous darts'.[14]

As with St Elmo's there is much parochial evidence of electrical storm involvement in 'ball lightning'. TORRO, the Tornado and Storm Research Organisation, received reports of several ball lightning episodes during the wet summer weather of 2007: one was at Denbigh, in North Wales, which damaged a phone system, and a witness saw a ball of light pass by outside a factory wall before striking with a loud explosion. A fireball in July 1992 made its appearance at a home in Dormansland, Surrey. Loud storm-driven explosions were heard and the house lights went out. Witnesses outside the house saw a huge red fiery ball of light (a 'BOL') plunge out of the clouds and burst into flames on the roof, setting the bedroom ablaze. All the electrical devices in the house were burned out.[15] In January 2007 a young German couple saw a white BOL fly through the window of their attic flat during a severe thunderstorm in the south-west of the country. It was accompanied by a hissing sound and a sulphurous smell, and the man felt static electricity and heat on his face – possibly attracted by the gold chain he was wearing. The light expanded, turned blue and exploded with a bright flash and report, scorching the man's face. It left a burn mark on the laminate floor, and a computer and TV set were damaged. On the ground floor of the building, two other people heard the explosion and saw the luminous ball drop from the ceiling near a light fitting, and when it fell to the floor it disappeared.[16]

In November 1979 a man called Robert Taylor, in an incident at Dechmont Woods, near Livingstone, West Lothian, saw a form of 'black ball lightning'. He then suffered an epileptic fit, which one UFO writer said had 'somehow induced a subsequent hallucination which took the form of a "circular flying saucer"'. The image of Venus at night 'was slightly above the nearby Deer Hill', and apparently induced the seizure.[17]

Many stories of BOLs, if they are not classed as St Elmo's, UFOs or spectral lights (see below), are more difficult to explain. In November 1902 swarms of 'fireballs' erupted across Victoria State, in Australia, during a fierce dust storm. 'Great balls of fire fell in the paddocks and on the streets, throwing up showers of sparks as they struck the Earth', reported *The Melbourne Leader*. Houses and stables were set ablaze, and timber props in coal mines were set on fire. Intriguing events were reported by a Romanian biologist, Alexendru Sift (1936–93). Over the years, BOLs and LITS became more intense both in frequency and variety, with a key date being 18 August 1968 when a disc-shaped object was filmed flying over the woods. Unidentified aerial lights seen in the Hoia-Bacu woods near Cluj, Romania, led researchers to say that this area is perhaps 'the most important area of unconventional phenomena in the world'.[18]

There were hundreds of Chinese reports during 1981 to 1985, when coloured lights were seen both at night and during daylight hours hovering just below the summits and ridges of surrounding mountains. Witnesses reported three types: a yellow 'bullet' with a sharp end pointing down; a strong blue-white light, sometimes flashing and always moving, and both types suddenly appearing and disappearing 'like flash bulbs going off'. Sometimes the blue lights formed a pattern of many light sources of different colours.[19] Some of the third type were 'Christmas tree' shaped.

A Bengali TV channel showed a fireball changing its shape from round to triangular and then into a straight line. It formed a halo with a range of colours, as can be seen from the footage reproduced in *UFO Data* magazine.[20] In another case, images of an illuminated UFO or fireball seen in the sky at night on 30 October 2007 in Kolkata (Calcutta) could be an optical illusion as a result of clouds, or digital fakery. Nevertheless, the UFO images were shown on Indian television, and commented upon by Debprosad Duari of the M. P. Birla Planetarium, who added that 'it does not appear to be a natural phenomenon'.

Notable BOL events

1659 At Markfield, in Leicestershire, 'terrible' claps of thunder were heard in a rainless sky, accompanied by hailstones of an unusual shape. Bangs like 'muskets' were heard, and 'prodigious eruptions of fire' were seen close to the ground. Some researchers relate these to 'earthlights'.[21]

1819 Fireballs, plus soap like sooty rain, darkened the skies over eastern Canada and New England on 10 November, during thunderous storms. Fireballs were seen playing around the cross of the French Paris church. The cross fell and broke into pieces.[22]

1847 Loud explosions and a 'barrel-sized flaming body' crashed to the ground at Forest Hill, Arkansas, on 8th December. The 'hot and sulphurous' object left a crater some 8 feet deep and 2 feet wide. It seemed to come from a black cloud that appeared suddenly, which had a red glare above it.[23]

1871 At Wilmington, North Carolina, on 18 July, a light followed by 'a hissing noise like a fire roaring', was heard and seen. The BOL then became 'three large stars of crimson fire'. A loud report was 'followed by a roll too long for a gun and not quite long enough for thunder'.[24]

1903 In May, at the genteel resort of Budleigh Salterton on the south Devon coast during a thunderstorm, lightning struck a cottage, and a BOL was seen.[25]

1938 In Northern Lapland, July, an estimated 30 fireballs drifted down from an overcast sky setting fire to anything combustible. Several houses were destroyed and people were badly burned.[26]

1975 In Barmouth, Leicesterhire, in December, a 23-foot fireball 'bounced down a street', crashed into a house and exploded.[27]

Are BOLs electrical effects?

In many obvious ways BOLs are *not* connected with lightning. Peter Van Doorn, director of TORRO, has studied the subject for over ten years, and he calls them Globular Light-emitting Objects (GLOs). This is a more appropriate name, he says, because they roll along the ground, and seem to have core features: 'There is no doubt that the GLO has a nucleus and that within that nucleus is a subatomic particle which can build itself into a body in an instant, disperse that body at will, and recreate it over and over again.' UFO researcher Albert Budden has theorized that many sightings of LITS and BOLs are the direct result of electrical effects, including electrical pollution from human sources. Ambient EM smog acts as a source of energy which can sustain a fireball in the atmosphere, he wrote.[28]

In November 1940 in Coventry, in the garden in clear weather, a BOL, about 2 feet in diameter and a pale turquoise colour, was 'a mass of writhing strings of light, about a quarter of an inch in diameter'. It flew over rooftops and damaged a pub about a quarter of a mile away. There was, however, an apparent occultic nature to this sighting, because the witness said before he saw the BOL he seemed to be plunged into the 'centre of intense blackness'.[29]

An important clue relates to the electrical effect on infrastructure – similar to lightning strikes. Some BOLs have registered pulses on magnetometers which suggest a strong link with magnetism of some kind. A BOL could also set up resonant frequencies, and could enhance an already charged atmosphere within a house with static electricity.[30]

Balls of light have an odd attraction to our telecom infrastructure. A ball lightning case was investigated near the small town of Sandy Hook, near Newtown, Connecticut in 1978. What was described as a bright, flickering rainbow was seen by a motorist, rotating and shooting out flares in all directions. The fireball then split into three smaller balls that circled his car.[31]

A glowing red shape about 200 feet in diameter got tangled with the power lines in East Anglia near Newark in October 1996. It was soon chased by an RAF Tornado flying dangerously low, although the aircraft itself seemed to be the familiar 'flying triangle' so often reported, with green and white lights in a chevron formation.[32]

A unusual aerial phenomenon around a television mast in the Manchester area in September 1983 was described by witnesses as a ball of brilliant orange light approaching slowly from the north-east. Other television masts in the area attracted faster-moving lights, some pulsating and changing colour.[33] 'Ball lightning' in November 1977 was witnessed by a coastguard in the Pembrokeshire Coast National Park. It was a very large yellow-green transparent ball with a fuzzy outline, and descended from the base of a 'towering' cloud and floated down a hillside. The coast-guard got static on his radio receiver while the object rotated slowly around a horizontal axis and bounced off projections on the ground.[34]

Some BOLs emit radiation EM, and eventually 'burn out', and seem to be associated with planes and cars because of the EM frequencies that arise from their ignitions and electrical systems. This could explain why some motorists have seen green luminous balls in front of their cars. A teenager in August 2006 reported being chased in her car by tiny floating lights as she drove across the Orwell Bridge on the outskirts of Ipswich. Two spinning lights appeared about 12 feet above the windscreen, each about the size of an orange, and gave off a diffuse, dull light. On the same night, similar mysterious lights were seen by other eyewitnesses over a mile away.[35]

A woman once drove into a bank of mist on a country road, to be confronted with a glowing, spinning ball of translucent greenish light about the size of a football. It stuck tenaciously alongside the car, about one foot from the side window, and kept pace even as the car gained speed.[36]

In a postwar incident a USAF cargo plane was flying over the Pacific when the crew saw two glowing curved horns of light projecting from the front cone of the aircraft. This corona was created by a powerful electric field generated by static electricity as the plane flew. Soon afterwards a light appeared in the cockpit – 'about the size of a volleyball. It touched nothing and made no sound', according to the pilot, Don Smith. The ball floated slowly down the aisle, went through a door, and exited out of the rear of the aircraft without leaving any trace of burn marks. Its appearance in an aircraft suggests a connection with gas discharged in powerful electrical fields generated by the plane's static.[37] In another similar case, an aircraft engineer and a pilot had stripped down a plane to its fuselage when a luminous ball of blue light entered the plane through the hull to the tail, and travelled back and forth a few times.[38]

BOL antics in buildings

It seems that ball lightning is more usually described as an indoor phenomenon rather than an atmospheric one, or at least one that frequently penetrates or bursts into houses and residences. Some BOLs – the more energetic ones – can penetrate walls and glass windows. One witness saw a glowing ball burn through the screen door of a house in Oregon, creep along to the basement, and wrecked an old mangle.

Physicist professor Boris Smirnov of the Russian Academy of Sciences in the 1990s sifted through thousands of eyewitness reports, looking for common denominators. The typical ball, he concluded, was reduced to a common image of an orange-yellow light around 8 inches across. It typically emerges from a thunderbolt or lightning conductor, floating for less than twenty seconds before vanishing, usually in a loud report.

Some scientists say BOLs seem to float for less than ten seconds before bursting when they hit a wall. Others say they can move at speeds of up to 33 feet per second, and hover about 3 feet above the ground. They can drift slowly or lurch and zoom downwards. Some explode or 'vanish' with a loud bang, and others just fade away. A woman in Yorkshire told how from her bedroom window she had watched in fascination as what she described as a 'dancing flame' appeared over the doorknob of the house opposite.[39] In another case an orb bounced on a Russian teacher's head more than ten times before vanishing. On New Year's Eve of 2006, a storm in Devon produced an explosion. A woman saw a brilliant, glowing ball, falling towards her house: 'It was so bright

I couldn't see the houses opposite, and it seemed to get bigger as it came towards the ground.' But there was no impact: the light had just vanished.[40]

Selected interior BOL sightings

1638 On 21 October, a 'thundering' fireball entered a church in Dartmoor, unpleasantly fiery and sulphurous. It crashed around the church, terrifying the worshippers, blasted through the roof and tore through a wall, ultimately causing many of the roof supports to collapse. Several people were killed and some sixty were injured.[41]

1902 A BOL event ripped up the ceiling, smashed crockery and left a sulphurous smell in several rooms. Fireballs were seen in November in both Australia and New Zealand over a sixty-day period, causing fires and explosions.[42]

1953 In the spring in a suburb of Washington DC, during a storm, conversation was drowned out by a loud sizzling from a patio. A point of light at a keyhole beamed a rod of light, then became a BOL of 10 inches in diameter. It flew over the heads of the people present and exploded against a brick fireplace.[43]

1962 In January, in Cheltenham, Australia, a glowing fireball passed a woman on her staircase and entered the bedroom, where it exploded harmlessly.[44] Almost at the same time a fireball came through an open window in Long Island, New York, and rolled across the floor between the husband and wife.

1963 A BOL emerged from a pilot's cabin and floated down the aisle of an Eastern Airlines plane on a New York to Washington flight on 19 March, just after lightning had struck the plane. One passenger described its 'almost solid appearance'.[45]

1975 A BOL emerged after a thunderstorm at Smethwick, in the Midlands. It was described as being purple-blue, and was 4 inches wide. It appeared above a stove in a kitchen. The woman in the kitchen felt its heat when she brushed it away, and it burned a 4-inch hole in her dress and tights, before disappearing with a bang.[46]

1977 In August in a house in High Barnet, two balls of green light no bigger than tennis balls were seen in the early morning as the sun was rising. During the following two weeks there were five electrical fires in the house.[47]

1979 A woman at Rowley Regis, in the West Midlands, in January 1979 witnessed glowing orange spheres hovering in her back garden.[48]

1991 In November, at a family home in Devon, strange glows seen in a room coalesced into a BOL, and then vanished with a 'pop' into a funnel of light.[49]

1992 Thunderstorm-related fireballs afflicted a family in Dormansland, Sussex, on 20 July. A loud explosion was heard and the lights went out, after which a 'small ball of red light' shot across the darkened living room. Other people in the area also reported weird fireball blasting, roofs catching fire, and electrical shorting.[50]

1996 Workers at a printing works in Tewkesbury, Gloucestershire, watched as a bluish fireball entered the premises on 7 June. It bounced along the ceiling, whizzed round girders and struck machinery, sending sparks flying. It oddly managed to pass through wire netting without losing its shape. This fireball was unusually vivid and explosive, since a loud bang accompanied by a massive flash of orange light occurred after it hit a window.[51]

2004 There were seven reports of ball lightning in the very wet August of that year, and they were especially seen in Beckenham, Kent. Two women saw a 'thin sword of blue-white light' pass through their kitchen, followed by a bang that knocked out the fuse box.[52]

2005 'Meteors' seemed to burst into flame and a giant ball of purple light was seen to crash through the roof of a house in Worthing, Sussex. There were also 'fantastic' noctilucent clouds.[53]

2007 During a storm in West Lothian on 1 July, a bolt hit a street with a deafening crack of thunder, shaking nearby houses, and producing a ball of 'glowing energy' that appeared over a computer in one family's living room.[54]

2008 A huge red ball seen hanging in the air from the window of a flat in Witley, near Godalming, Surrey, on 31 August was a very unusual sighting because of the size and clarity of the object, said to be 'perfectly circular'. It was the length of a London bus, and 'unbelievably vivid'. 'It was like a ball of fire, but with no flames, no edges to it, and it didn't move', according to Bill Brooks, an eyewitness. Soon after there was a blinding flash, and a huge crack of thunder shook the building.[55]

2009 Ball lightning, described as a 'glowing ball of orange fiery light', was reported outside Longside, in Aberdeenshire, on the evening of 26 August.[56]

The 'plasma ball' theory

As usual, there remained enigmas that challenged university physicists: what do the laws of buoyancy say that could possibly keep the smaller indoor BOLs afloat? They must surely have a density no heavier than air to do this? The fact that *some* BOLs are cool, according to the record, or at least just 'warm' – rather than hot (but not searingly hot) – perhaps explains why they hug the ground or are observed at room level. Older reports often said BOLs gave off little heat, but this is now generally discounted. (The daughter of King Henry II of France described how a glowing object entered her bedchamber via the window and floated around the room, singeing the bedclothes.)

On the other hand, a ball of incandescent matter would struggle to remain spherical, and would do a lot more than singe clothes. Even the idea of a plasma ball bound together by some form of cohesive energy field is difficult to comprehend. How could magnetic fields or EM waves follow the ball as it floated? If plasma balls are 'cool', some scientists believe they would have to travel higher into the air to enable them to keep their round shape. On the other hand, if the physics is correct, they couldn't rise too far because they would need to become coated with successive spherical layers created by different chemical processes. This means that one of the layers would be laden with water droplets to create a cooling effect and provide a surface tension. This would not only shape it into a ball, but keep the heat within the ball. Perversely, however, this concentration of moisture would weigh the ball down, counteracting the lift of the hot gases inside.[57]

A scientific consensus is slowly emerging around a scarcely understood phenomenon. 'Plasma balls', a peculiar interaction of cosmic rays with the Earth's electrical field, or some kind of charged-up vortex, may be involved. A large number of BOL sightings occurred in early 1977 in Lancashire and the Pennines, accompanied by high-pitched noises.[58] One scientist said the most likely hotspots for 'plasma balls' are Warminster in Wiltshire, Shoreham-by-Sea in West Sussex, Brecon in Wales, and Guernsey. Indeed, several photos of 'ball plasma', all with reddish outlines (and 'double-jointed'), taken over the Vale of Pewsey, Wiltshire, in 2001 and the summer of 2003 were shown recently in *UFO Magazine*. In fact, hundreds were seen during the 1990s and early 2000s in that region.[59]

Scientists sometimes bring up the silicon theory of ball lightning. John Abrahamson of the University of Canterbury, in New Zealand, and James Dinniss said that Earth's silicon-rich crust is unstable at high temperatures, and likely to be 'activated' during lightning strikes. When a thunderbolt strikes the ground, a chemical reaction with silicon can cause a hollowing effect in the ground. A vapour made of hot, silicon-rich particles rises to form *fulgarite*, (see Chapter 5), then cools. Particles are then drawn together by the silicon that has been charged up, to form a ball of 'nanostrings' which start to glow as the silicon reacts with the oxygen. Organic factors in BOLs can be due to oxygen that can also react with other gases to form nitrobenzene, which is an oily substance with a strong smell rather like bitter almonds.[60]

Dr Gerson Paiva of the Federal University of Pernambuco, Brazil, described how he had blasted samples of pure silicon with electric arcs in the laboratory to test the idea that BOLs turn solid silicon into silicon vapour. He and his colleagues took wafers of silicon just 450 micrometres thick, and zapped them with currents of up to 140 amps after placing them between two electrodes. By moving the electrodes slowly apart, and creating an electric arc, they caused bits of fragmented silicon, blue-white or orange-white in hue, to be thrown out. Some, indeed, became luminous orbs the size of ping-pong balls, but these only lasted some eight seconds.[61] However, none of them floated, even though they did display occultic behaviour, producing smoke trails, and with fuzzy surfaces on the balls from which little jets seemed to emerge, jerking them forward or sideways.

Some fifty years ago, while on a camping trip in Netanya, Israel, Dora Jerby and her friend saw a football-sized glowing ball by the seashore. 'It seemed to be emitting light from its core', she said to reporters. Her scientist son, Eli Jerby, had developed a microwave drill at Tel Aviv University to create molten hotspots on the surface of a small rock. These molten blobs would sometimes burst into a floating ball of flame.[62] Unfortunately, however, even with the microwave source being permanently on, they lasted only for milliseconds, hardly long enough to be seen by the human eye.

The 'earthlight' syndrome

Eyewitnesses throughout history have seen strange lights prior to an earthquake. Immanuel Kant seemed to be aware of the connection when he wrote in regard to the 1755 earthquake in Lisbon: 'Of several other earthquakes, violent lightning in the air and the fear that one notices in animals have been the precursors.' Early examples of these peculiar lights came from mountaineers in the nineteenth century who reported vibrating and ringing resonances, and of mountain rock 'bursting with glowing electrical currents'.[63] Strange landscape lights even inspired a 1999 episode in the television series The X-Files.

Earthlights were cited in England in 1750, and were described as 'reddish bows in the air' after a tremor was felt on 2 March.[64] Luminous globes were seen on the day of an earthquake in Boulogne in 1779,[65] and earthquakes were cited as early as the 1890s as an explanation for a fireball phenomenon seen in a house in Perthshire.

On 26 November 1930, Izu peninsula in Japan experienced over 1,500 reports of earthlights. Some were centred on the Tokyo Bay area, and the lights there were described as either 'auroral streamers' or 'fireballs, and usually bluish'. Other reports added that when the earthquake was at its height, a straight row of round masses of light appeared in the south-west.[66]

Strange lights were seen around a Chinese mountain called Wu T'ai Shan in the 1930s. Explorers and geologists had noted these 'mountain peak discharges' before. A man called John Blofeld saw 'innumerable balls of fire [that] floated majestically past ...' the mountain. These lights were described as being beams of light around the peak, or a glow that sometimes gave way to bursts of 'searchlight' beams. They have been noticed in the Chilean Andes, and are sometimes visible over huge distances.[67]

In 1932 earthlights were seen after a minor quake in Humboldt county in California. The lights appeared as 'bolts of lightning travelling from the ground towards the sky'. In Britain, lights were also seen just before a strong quake in 1957, and centred on Charnwood Forest, Leicestershire. They appeared as 'tadpole shaped', and flew about. In fact, considerable research was done in the 1970s in the earth-tremor epicentre of the county of Leicestershire by researchers Andrew York and Paul Devereaux, and by other local historians aided by archives and local newspaper interest.[68]

The longer frequencies of light often seemed to occur near quakes. In eastern Tibet in 1947 a large quake was followed by vivid orange-red glows, and spherical light-forms seemed to hang for days over the eastern sky.[69] In 1968 the first photos of 'earthquake lights' were taken by Yutaka Yasui of the Kakioka Magnetic Observatory during a series of quakes at Matsushiro, Japan. Some showed red streaks across the sky, like low-lying aurora. From a distance others looked like a low 'blue-coloured' dawn. In November 1979 'yellowish' lights in the sky over Lancashire and Merseyside were described as earthlights by Professor Peter Sammon of University College, London.

Three clergymen at Cefn Mawr, inland from the Barmouth–Tywyn area of Wales, saw balls of light 'rise from the earth and suddenly burst luridly'. These sightings overlap with the earthlight syndrome because the area was known to experience tremors in the past. In fact, northern coastal Wales has a record of people seeing either BOLs or earthlights. In July 2005 strange lights again appeared around Barmouth. Dozens of reports continued for months of lights rising from the ground, often hovering. Most were spheres, but there were other angular shapes. There was unusual seismic activity at the time, and two years earlier a small quake had struck nearby, and a large ball of fire was seen to rise from the ground and suddenly burst. Luminous multi-coloured objects were seen in skies over Tepexpan, Mexico, on 26 November 2007. These were filmed and shown in the *UFO Data* magazine of January–February 2008. They remained static, and could have been 'earthlights' because they coincided with a strong tremor in Mexico City.

Momentary light phenomena can be explained by physics. Many dry substances, like sugar, give off light by having their electrons stripped away by the applied force that snaps bonds between atoms and even between subatomic particles, where the electrons recombine with atoms. Even crunching mints in a darkened room will give off bluish-white sparks because the tiny voltages that are generated excites the nitrogen gas in the air and makes the gas luminescent.

Many scientists, like seismologist Dr Roger Musson of the British Geological Survey, agree that the quake lights are real[70] (although Dr Ian Griffin, a former Nasa astronomer, says most lights are merely space debris).[71] US geologist John Derr built an impressive body of data on the 'tectonic strain' theory based on earlier *sonoluminescence* (sound and light) theories – the mysterious flashes of light that occur when bubbles of any liquid are blasted with powerful sound waves. High and low-pressure can build up in any liquid, but bubbles seem to emerge from very low-pressure situations, and when the bubbles collapse – probably involving photons and other subatomic particles in the water molecules – they emit light. This might be called a *micro* theory of light emissions, whereas an original tectonic earthlight theory published in 1977 by Canadian scientists Michael Persinger and Ghislaine Lanfreniere could be described as a macro theory. Five years later Paul Devereux and the geochemist Paul McCartney published a seminal book, entitled *Earth Lights*, that elaborated on this macro theory (of earth substances rather than gas particles). Lights often appear over geological fault lines during periods of tectonic activity as rock layers grind against one another, and this results in the well-known *piezoelectric* and *piezomagnetic* effects.

But Devereux has been accused of pushing naturalistic explanations for LITS too far. He cites, for example, the case of a couple standing in their back garden in the mid-1960s in Bridgend, Wales, when they saw a 'sparkling mist' appear low on the horizon, accompanied by what they thought was the sound of a distant jet engine. Clearly, however, Devereux was having a conventional UFO story related to him, with the mist glowing, pulsating, rotating and dividing into two sections.[72]

Nevertheless, piezo lights were associated with the Tangshan earthquake in 1976, in which 240,000 Chinese lost their lives and their city. For several days prior to the great Kobe quake in January 1995 in Japan, there were again reports of glowing orange-red and pink lights hovering over the Kobe faultline.

The gist of the earthlight theory is that the inner core of the Earth is shaken up by seismic activity which in a sense 'super-charges' it, so that giant electrical sparks are released from the core via vents in the Earth. The earthlight effect seems to be one of spatially enlarging and magnifying the geomagnetic field, or another field not sufficiently understood that exists at ground level.[73]

Earthquake lights seem to cover the entire EM spectrum, but are more often associated with Extremely Low Frequency (ELF) signals. A Spanish episode of UFOs in the late 1960s was also said to be connected with faultlines at a time when daily earth tremors were being experienced. Loud detonations were heard and 'immense' blue-green fiery balls were seen, along with conical and 'searchlight'-type lights.[74] Research by the Michigan Anomaly Bureau suggests that UFOs in the state were connected with earth-lights and faultlines.[75]

The earthlights themselves could also be carried along by the atmosphere in one direction, assisted often by ambient radio waves.[76] 'Unusual radio waves' had also seemed to occur prior to many earthquakes, hinting that such 'harmonic energy fields' could be enhanced by human activity.[77] In March 1992 the *Washington Times* referred to ground sensors that detected mysterious radio waves or 'related electrical magnetic activity' before major quakes in Southern California in 1986–7, Armenia in 1988, and Japan and Northern California in 1989.[78]

But it is not known whether the molten core of the Earth, some 1,240 miles in diameter, conducts electrical flows, and whether this could account for an additional electrical field that could excite gas molecules. In 1981 Brian Brady of the US Bureau of Mines filmed an experiment involving massive pressure applied to a slab of granite. True, small BOLs did seem to emerge. Rock is known to be a good conductor of electricity, and scientists at the USGS, some years ago, said rock friction could also generate a 'sheath' of steam. The moisture released from the Earth then acts in the same way as a containing envelope for BOLs.[79]

Nevin Bryant, a remote-sensing expert at Nasa's JPL in Pasadena, proposed a theory that decreed that igneous rocks, which normally act as insulators, briefly behave like semiconductors, and from time to time the rocks emit positive electrical charges. Positive charges inside the Earth can operate in more than one way. *Crystals* in volcanic rocks contain paired oxygen atoms, called peroxy groups, which can snap under stress. This releases negative ions of oxygen which remain trapped inside the rock strata, while a positive charge is released to flow through it. Since like charges repel, all the positive ones will swiftly flow outwards, rather than remain in the rock.

The visible spectrum is brought into play when the positive charges spread out on the surface in a thin layer. The field then becomes huge, around 400 kV per centimetre. This in turn – the charged-up field – would ionize the air up to 12 feet above the ground, causing luminous plasma to become visible in certain frequencies. So electrical activity, combined with seismic activity, could cause flashing lights to appear above the surface. The silicon theory could come into play. Possibly igneous earth rocks free pure silicon from the soil and this glows as it reacts with oxygen in the air.[80] In late April 2007 a pilot reported bright yellow lights shaped like flat discs in the sky near Guernsey. Paul Devereux (probably wrongly) pointed out that it may have been a precursor for a quake that struck off the coast of Dover. And on 31 March 1982, EM emissions were recorded on sensitive monitoring instruments about half an hour before an earthquake struck on the Idu peninsula of Japan.[81]

A question mark has arisen as to whether earthlight charges would break out of the surface of the Earth instead of dissipating inside it. Earth's internal moisture and liquids are regarded ambiguously. Earth's crust is saturated with water, which should quench the build-up of a charge of some sort. On the other hand, water could be a conduit that helps transmit the quake tremors and associated earthlights further along the fault. (Earthlights, incidentally, are unlikely to form over water.)[82]

In many earthquake events clouds of toxic ash and dust tower up and ionize the air, generating an explosive electrical storm around the dust plumes. Eventually an entire cloud turns into a colossal battery, generating and deploying kinetic and electrical energy in large quantities. In Chile a volcanic eruption met with a lightning storm in May 2008, and tons of dust and ash from the eruption of the Chaiten volcano poured into the night sky.[83] Volcanic fireballs are probably the single most destructive weapon in nature's armoury, short of black hole or antimatter impacts. Pyroclastic flows – a devastating super-eruption – are also called *nuees ardentes* – or 'burning clouds'. Hot gases may account for the sulphurous smell that is often reported.

Stray 'earthlight' voltages

A US version of earthlights might explain why hundreds of dairy farmers across Wisconsin and sometimes also in other states still struggle with 'stray voltage'. The electrical fields, piezo effects and radio waves discussed above could have caused cattle to die or become crippled, and a group of farmers presented a list of demands to an energy cooperative board of directors in Arcadia, Wisconsin. Concerns were increasingly expressed in the US business and scientific community. Duane Dahlberg, a retired physicist and former consultant for the Electromagnetic Research Foundation in Moorhead, Minnesota, had studied the growing significance of ground currents since 1983. In 2000 he pointed to the increasing load in the US.[84] Further, the Electric Power Research Institute in Washington State and in Palo Alto, California, said 70 per cent of all electricity used passes through equipment that produces wavelength distortion – called harmonics. It is this that puts heavy loads on the grid – up 30 per cent in the 2000s.

A village in Sicily, Canneto di Caronia, suffered puzzling electrical failures, unexplained fires and smoky outbursts that struck nine houses, displacing seventeen families. Televisions and fuse boxes blew, kitchen appliances shorted out loudly, and fires started spontaneously throughout 2003 and 2004. Scientists could not understand why officials from several agencies, like the National Institute of Geophysics and Volcanology of Bologna, and the

National Research Centre concerned with building and earth materials, had barely a clue as to why the electrical failures had happened. Lay theories and explanations abounded. Perhaps, went one theory, supercharged positive effects arising from beneath the crust in the northern region of Sicily caused electrical energy to seep into the village. Perhaps volcanic activity could have caused sparks to fly, especially as the hamlet was near transmission lines and electrified railroad tracks.[85]

Theories about the earth acting as an electrical conductor were also voiced by Dave Stetzer, an electrical contractor from Blair, Wisconsin.[86] The science of electromagnetic compatibility, or EMC, has prompted attempts by scientists to overcome the power outage problem by desperately trying to jam or countermand the negative effects. Inevitably, these fail and many victims of these events try other methods. In September 2004 many residents in Caronia evacuated their homes after calling in an exorcist. After rewiring by the country's major electrical company, Enel, the power outages and fires ended.

7 Folklore Lights and Mirages

There are basically three types of UAP applicable to this chapter: the gaseous *Jack-o'-Lantern* type; the spectral lights that have folklore legends surrounding them; and the mirage that gives rise to all sorts of illuminated apparitions. There is one important distinction to be going on with though: any lights seen in the sky or over the horizon from only one angle have to have a different explanation from lights that can be seen by other unconnected witnesses at different locations, and lights that move around are different from stationary phosphorescent lights, although the latter often seem to flicker.

Fleeting Earth-bound lights have been known since the time of the ancient Greeks. Dark Ages chroniclers referred to them as 'fire dragons', while the sixth-century AD Gregory of Tours called them 'miraculous acts of God'. They have often been seen over long, low mountain ridges. A German engineer, John William Gerard de Brahm, recorded mysterious mountain lights in 1771. He suggested they were probably 'vapours'.

Lights described as St Elmo's and mentioned in the previous chapter, normally associated with lightning, apparently occurred in 1923 in the village of Feny Compton, in Warwickshire. Crowds watched in awe as the lights danced around for nearly a year at periodic intervals. Sometimes they hovered near a deserted farmhouse.[1] Local newspaper reports of this event may have simply used St Elmo's in a generic sense. While there is no clear link between BOLs and St Elmo's, there is even less of a link with the similar will-o'-the-wisp, which is a phosphorescent light, also known as the Jack-o'-Lantern. At Alderfen Broad, in Norfolk, a Jack-o'-Lantern was often seen in the early 1800s, twisting and turning. Legend had it that it was possibly the ghost of a murderer who had drowned there.[2] In South Yorkshire, the luminoscities are known as '*Peggy wi't lantern*'. In Wales they are known as 'corpse handles'. In German folklore they are known as *Dickenpoten*, or as *Irrlichtern*, meaning a 'wandering light'.

Organic lights like these perhaps drift out of the UV spectrum and produce light from natural surfaces that fluoresce and become observable when other wavelengths of light are low. Sheep sometimes seem to glow at twilight, as do some flowers. These examples typically occur when static in the air reaches a high level and energizes tipped or pointed objects. Cattle were once seen standing in a pasture at night with their horns glowing silver and blue, although animals seem unaffected by the glows.

'Marsh gas' (sometimes known as 'swamp gas') arising from eruptions of methane or phosphorus hydride, or similar gases, can also be classed as a will-o'-the-wisp. As the flames form and die they appear to flit around boggy ground. Diphosphane is another marshy and a highly flammable gas that is said to burst up from rotting vegetation to become ignited. But this would be a kind of spontaneous combustion, and the descriptions of it don't follow a typical pattern for these gases, emitting no heat and moving freely in any direction. Further, phosphorus is never found in a pure state in nature, and laboratory tests have so far been able to detect only insignificant amounts of phosphorus in marsh gas anyway[3] (and it is seldom explained where the spark that ignites the gas comes from).

'Spectral' lights

Ignis fatuus means 'foolish fire' in Latin. It is regarded as a mischievous 'spook', or spectral, light that attracts travellers often to their doom, but just as often leads them to safety. Collectively, these eights appear as blue or yellow flames or globes, floating just above ground level.

The USA has a rich heritage of 'spook lights'. Many of them in some way seemed to possess an awareness and must be excluded from other 'organic' lights, although many 'marsh gas' lights, known by locals, have occultic legends attached to them. There are cases where this clearly happens, and we must include in this category lights that are often seen in the road about human height.[4] They are often, if not invariably, associated with spirit legends of tragedy, death and drowning in local bogs.[5] Early frontiersmen believed that the lights were the spirits of Cherokee and Catawba warriors.

There are literally hundreds of sites worldwide where spectral lights have been seen. They are said to appear in the homes of the dying and are seen floating down country lanes at night, as they did in Barmouth in 1905.[6] In 1923–4 at Burton Dassett, in Warwickshire, rumours of UFOs or 'ghost lights' abounded. A yellow and blue light or ray was said to flit over a lonely range of hills in the area, indulging in typical UFO-like changes in colour frequencies. A motorcyclist told of how a brilliant light hurtled down the road towards him, and people attributed this to a form of Jack-o'-Lantern, or ghost light. A woman claimed to have seen a yellow-blue flame bobbing among derelict farm buildings before suddenly turning into a dazzling glare.[7]

Early last century two Lake District fell walkers, returning to Keswick, passed by the Castlerigg stone circle just in time to see white luminous spheres floating around the circle. The BOLs winked out suddenly as the walkers, curious, tried to approach them.[8] During the Second World War British troops stationed on the Island of Curaçao, in the southern Caribbean, and marching at night, saw a descending green light about 2 feet in diameter. It bounced along the road, and one officer quickly ordered some men to catch it, but it evaded them and shot off the road into a culvert.[9] These spectral lights also seem to instinctively avoid cars or people, and 'blink out' when approached, only to reappear much farther away. They were investigated by US army engineers during the war, but the results were inevitably inconclusive.

The *Marfa Lights*, which come on only at night and in most seasons, are often centred around the Chinati Mountains, and can often be observed near US routes 90 and 67 on Mitchell Flat, between Marfa and Alpine, Texas. Through binoculars, they can be seen as small dots that enlarge to become a glowing white ball. They often blink on and off, fade or pulsate. Appearing between ten and twenty times a year, the BOLs can reach the size of basketballs, and usually float just above the ground, although they can go higher. They are 'multi-coloured', and each one is different.

They can appear either in singles, or in pairs or in groups, and can last seconds or hours. They were known to the Apache Indians who said they were Indian spirits. A team of geologists went to make observations of the Marfa Lights in 1973. They saw the lights, but were unable to get close enough to draw any valid conclusions. They saw them swinging in the air, rocking back and forth and looping, like insects. The photos taken were of no real evidential value.[10]

Attention has focused on the famed case of the 'Lubbock Lights' over Lubbock, Texas, in August 1951. Four Texas Technical College engineering professors observed two strange formations like 'strings of beads in a crescent shape', but they were generally regarded as UFOs. One of the observers admitted he would have made no comment about the sightings if he had not been backed up by other academics. The lights were turquoise, and there were two or three flights of these objects per evening observed over several days, with a varied number of fifteen to thirty separate lights on each event. The objects always appeared at an angle of about 45 degrees from the horizontal in the north and disappeared at about forty-five degrees in the south.[11]

At Gonzales, Louisiana, in April 1951 and Suffolk County, Virginia, the sheriff, Hickley Waguespack, was among the witnesses who watched a ball flitting over treetops and over the road. The Suffolk County lights, known in history to indigenous Indian tribes, on 5 March 1951 hovered over rural highways 5 feet from the ground, in particular on the Jackson Road. An investigating state trooper said that one light looked like a locomotive headlight.[12]

The Surrency, Georgia, spook light is occasionally seen along railroad tracks serving the small town with a population of about four hundred. It is beachball in size and bright yellow. During a seismic survey geologists from the University of Georgia in 1985, found a bizarre dome-shaped pocket of an unknown liquid at a depth of 7 miles. Some believe this is responsible for the Surrency light, but others say it is the ghost of someone killed on the railways in 1870. There have been poltergeist activity and mysterious noises in the area for years.[13]

In December 1995 a young couple drove along the Reservoir Road, near the mines in the Brewster area, not far from the River

Valley UFO sightings in the USA. The man thought he heard voices telling him to stop. He then found that his car was plagued with a sulphurous smell. When he drove off again after a short while the couple heard a loud bang and the car vibrated. Other explosions seemed to come from the woods on top of a nearby hill, when a ball of light emerged.[14]

Other notable US spectral lights

- Ghostly orbs were seen at Silver Cliff, Colorado, in April 1956. They were bluish and pulsating, the size of a basketball, sometimes appearing in twos and threes. A crowd pursued the lights around a cemetery one night, but the lights would go out when approached.[15]
- In North Carolina they are seen as the Maco Station lights, or the Brown Mountain light.[16]
- In Arkansas the Dover light is seen almost every night, according to newspaper reports. It is reputed to have animal-like instincts.[17]
- In Maryland there is the so-called Hebron light.[18]
- Lights are seen at the Spooksville Triangle, where Missouri, Oklahoma and Arkansas merge.[19]
- The Crossett spook lights in Arkansas are 'multi-coloured' and always keep their distance from encroachment. The local legend said this light was the ghost of a switchman killed in a train accident around 1850.[20]
- Spook lights have been seen stretching from El Paso southeastward along the Rio Grande valley, past Big Bend National Park and southweastward into Mexico.[21]
- The Hornet light, first seen as long ago as 1901, is said to be the spirits of Quapaw Indians who died many years ago, or perhaps the ghosts of murdered children.[22] It, or they, tend to glow in an orange hue, bobbing on summer nights, year after year, 12 miles south-west of Joplin, Missouri, and 2 miles south-west of Hornet village. They have often been photographed. In 1913 the *Charlotte Observer* described them as 'much smaller than the full moon, much larger than any star and fiery red'.

'Spook lights' around the world

The writer Mary Kingsley on a visit to Gabon in 1895 saw a ball of violet light roll out of a wood on to the banks of Lake Gomboue, where it was joined by another. Giving chase in a canoe, her friends saw one of the balls fly back into the trees while another dived below the water, glowing as it sank. There are also the '*Naga*' fireballs of Thailand, seen every October emerging from the Mekong River. Hundreds of them have been recorded along the 80-mile stretch of water, and thousands of visitors come to see them each year. Local legend says the fireballs are from the *Nak*, a serpent from Buddhist legend. The official explanation is that they are methane gas eruptions from the water, although some dispute this.[23]

The most famous spectral lights are to be found in Norway. In 1984 hundreds of witnesses could see bizarre LITS in Scandinavian skies almost every two hours around the clock. These lights persisted for most of the early 1980s, particularly in a 7-mile valley in Hessdalen, in central Norway, 74 miles south of Trondheim. Suggestions that the lights could be some form of piezoelectric earthlights were tested by the Project Hessdalen team, one of the most scientifically respectable investigative bodies delving into the mystery of these lights, which people were beginning to say were 'spectral'. Unfortunately, they had little success after many years of probing, waiting and watching with the latest electronic technology. And this despite the willing assistance from the Norwegian Defence Research Establishment and from the University of Oslo. Dr Odd-Gunnar-Roed and others monitored the area with radar systems, magnetometers, spectrum analysers and other instruments.[24] But no increase in background radiation was ever found, nor was any kind of electrical or magnetic field interference discernible, nor was heat detected from the lights.[25] It was only when a powerful laser was deployed (in other words, when a beam was *emitted*) to see if a strong light source would react to the lights, that something happened. In eight out of nine times, the LITS did indeed react by dimming, fading out or changing their pattern, hinting at some kind of autonomous or even sentient behaviour unconnected with natural atmospheric phenomena.

The lights tailed off after the 1980s, although they are still occasionally seen, and at a higher rate of prevalence than similar types of aerial phenomena elsewhere in the world. Research continued spasmodically in the 1990s through to 2001, when an Italian team joined the Norwegians. This team was headed by Dr Massimo Teodorani, an astrophysicist from Consiglio Nazionale delle Richerche in Bologna. Together the scientists formed the EMBLA Mission of 2002. In September 2006 the Norwegian/Italian scientists recorded yet again a floating light that couldn't be explained.[26] The Hessdalen researchers said the Norwegian lights were quite unlike the spectral lights seen in the USA. The mountainous northern terrain, they pointed out, tended to rule out any gaseous or will-o'-the-wisp type of explanation. Dr Teodorani believes the Hessdalen lights are thermal plasmas 'of unknown origin'.[27]

Mirages and apparitions

One puzzling aerial phenomenon (or 'UAP') is the way in which different air densities turn the atmosphere into a giant lens, in the same way that air densities also deflect sound waves (see Chapter 12). Thermal inversions – where warm and cool air layers swap places – occur where the air temperature increases (rather than decreases, which is what normally happens) with altitude, or where the water vapour content decreases with altitude. The result: layers of heat and air inversions twist light rays away from the far distance, making them appear distorted.

This phenomenon creates 'mirages'. Bizarre images of upturned ships can be seen, while entire towns can miraculously float off the sea. Far-off oceans or surface ice, which are essentially flat and horizontal, (i.e. they are on the horizon) sometimes appear to observers in the form of vertical columns and spires, or as 'castles in the air'. This also happens pretty quickly, too, with the shape of the apparition changing within less than half an hour.

Intriguing mirages have been experienced throughout history. A Roman garrison was mistakenly ordered to march to the coastal town of Ostia, 20 miles to the east of Rome, which, when his aides pointed out the red glow on the horizon, Iberius Caesar in AD 34 thought was on fire. In 1650 Cromwell's forces routed a Royalist army of Scots at Dunbar, East Lothian, after which 'the sky opened in a fearful manner and a terrible fiery shaking sword appeared with a blue handle'.[28] A mirage once appeared over a walled town at Youghal, Ireland, in October 1796, which was so precise that complete garden layouts could be seen.[29]

Arctic skies have often displayed scenes of fantastic castles and spires towering up over the horizon. In 1818 John Ross, an explorer, searching along the north coast of Canada, surprisingly found his way blocked by what seemed to be uncharted mountains, and decided to return home. The local Inuit referred to the vision as 'poojok', roughly meaning a mist. Ross had seen the *fata morgana*, which is another name for the atmospheric distortions of light giving rise to mirages, named after Morgan le Fay, the legendary half-sister of King Arthur who could supposedly fly and change shape.

Captain Robert Bartlett and his crew were on a trip to the Arctic in 1939. When his ship was midway on its journey just beyond Greenland, he was stunned to see the contours of Iceland appear a lot closer than they should have been.[30] This was the mirage effect caused by warm air thermal inversions, mentioned above, which bent the light waves from the horizon.

On rare occasions people living on high ground overlooking the coast of the Isle of Wight can see, with binoculars, house lights winking all the way from Cherbourg, some 60 miles distant on the coast of France. And tourists in Dover, from the White Cliffs, and without using binoculars, often see the refraction of the town of Dunkirk on a clear warm day with the breeze blowing in the right direction, although the distance is about 24 miles and the town would normally be below the horizon.

The way images reverse and re-orientate themselves is particularly unsettling. On the east coast of Scotland in July 1860 a warship was doubly reflected 'as if two ships were sailing one above the other' before one ship became 'inverted'. Trees were magnified 'into gigantic forms' and 'the houses distended into factory chimneys, with windows all down them'.[31]

In 1897 a correspondent to the weekly newspaper *The English Mechanic and World of Science* described how he had seen a 'city in the sky' over Alaska. A mirage over the Humber Estuary was reported near Grimsby in April 1909: 'Three miles of land …

appeared to be lifted high into the air and reversed, the trees inland having the appearance of growing upside down', reported *The Times*. There were steamers at sea, with mast and funnels, all seen reflected from miles away on to the land in an apparent response to the variegated warm and cool layers of the Humber.

A crew of a ship in the North Pacific in March 1937 saw a steamer apparently travelling upside-down. Then they saw another on the horizon appearing above the real one, with a widely distorted image of the previously upside-down steamer now joined to the real ship funnel-to-funnel. A short while later there appeared to be three ships sailing along the top of one aother, with the middle one inverted.[32]

In China in 1984 a mountain was seen with small houses at its base. Then these metamorphosed into a 'flatland' with two small hills flickering into view, before suddenly turning into shimmering high-rise buildings.[33] A Chinese mirage occurred in June 1988 when a small town was seen surrounded by hills, apparently floating. In the following month more scenes of mountains, lakes and buildings appeared, never to be seen again until 2006 when thousands watched as a mirage appeared over the sea near the Chinese port of Penglai, a peninsula on the north-east coast in Shandong Province. Then mist rose from the shoreline to create an image of a city with high-rise buildings and broad streets, which could have been a carbon copy of the city of Dalian, about 60 miles distant across the peninsula. It appeared as if a city was floating in the sky.[34]

A pilot on a local light aircraft flight from Southampton airport to Alderney in the Channel Islands on 23 April 2007 saw a yellowish light in the sky. He thought at first it was sunlight reflecting off garden greenhouse panes until he realized that the reflection was not disappearing. Binocular vision showed it was disc-shaped, and other passengers confirmed this sighting. The captain, Ray Bowyer, said it must have been a mile wide! Grant Allen, an atmospheric scientist who had analysed weather patterns for that day, reckoned that a temperature inversion could have created the illusion, although the pilot disputes this. (Another pilot said that what the first pilot saw was probably of military origin.)[35]

*

The 'Min Min' lights are found in Boulia, south-west of Queensland, and are named after the hotel from which they were first sighted. The *Min* is an aboriginal word, and the Min Min theory was given backing by Bill Physic of Australia's Bureau of Meteorology in Melbourne. These lights often appear as undulating orbs, bathing witnesses in a cold, whitish light. They seem to bob and weave, and motorists say they appear to lunge at them, or chase them at high speeds.

Jack Pettigrew and colleagues at the University of Queensland in Brisbane saw one Min Min as they drove across the outback, which appeared as if it were 100 yards away. It seemed to remain motionless even after they had travelled a further 3 miles. Checking maps and compasses they discovered that a convoy of trucks with their headlights blazing had crossed a section of road near a small community called Windorah that was some 180 miles away beyond a range of hills. Later experiments with truck headlights by Pettigrew beamed at a low elevation behind a hill were monitored by friends miles away as the Min Min blinked on and off in perfect sync.[36] Pettigrew concluded that the Min Min lights were a typical mirage, with the outback producing an inverted image, bending lights from car headlights or a bushfire some distance away. Hence distant images were seen, and seen from a different angle, and at a higher elevation, from their original source: in fact, car headlights could be seen from a distance of tens of miles. (On the evening of 8 August 2009, the occupants of a car near Bridge of Brown, Moray, in the Highlands, appeared to see red traffic lights on the opposite side of a valley.[37] This was said to be a Scottish version of a Min Min.)

Spook lights at sea

On the morning of 12 April 2009 the commander of the US guided missile destroyer USS *Milius*, while out in the Arabian Gulf, saw a strange light phenomenon in the ocean waters. The commander, Kendall Gennick, said he saw a wave of white light 'pulsating out of nowhere', and that was tinged with 'millions' of tiny green lights that appeared on the surface of the waves. In fact, the entire phenomenon was shrouded in a glowing and pulsating green mist.

Yet there was no reflection of the moon shining on rippling waves, because the commander said the night was 'moonless' and overcast. Another of the ship's officers pointed out the outward expanding circular waves, and said that the light 'spun and moved faster than the waves, and made circles and other patterns'.[38] The swirls of coloured lights splayed out like magnetic filings, as if energized by a giant undersea vortice. However, although the display stretched to the horizon, it didn't seem to have a central point to it – such as, for example, a strong beam of light from below the waves that could possibly be the work of scientists using submarine modules.

This phenomenon was unrecognized by the scientific community, but had been reported on earlier by the *Nexus* magazine, the journal of fringe science and suppressed discoveries, in 1997.[39] Later it became known – unofficially – as the 'Marine Lightwheel Phenomenon' because of the growing number of similar reports. An earlier event was dated to May 2007 in the Straight of Hormuz by another US navy ship, and seen by thirty crew members. Kendall Gennick thought there might be a conventional explanation for it, and he mentioned either St Elmo's fire or perhaps bioluminescence arising out of the stirring up of plankton-type species that had phosphorescent characteristics.

In the books by Richard Winer and Charles Berlitz about the *Bermuda Triangle* – that is, the area bounded by the Bahamas, the Straights of Florida and Cuba – we learn of strange disappearances of ships and aircraft at sea, although pilot or ship's captain error or mechanical failures are usually ruled out. Radio 'dead zones' appeared where messages could not be sent or received. Radar images of objects appeared and disappeared from screens, and bluish glows mysteriously flooded cockpits.

The Great Lakes of North America – another notorious 'disappearing' zone – are completely enclosed in freshwater pools, and it is reckoned to be impossible for any plane to be more than twenty minutes from land. Any pilot in difficulty in that region would still be able to glide to safety, and if the planes had crashed while doing so, there would be obvious traces of the wreckage. Further, there were hundreds of ground-based and sea-based radio monitoring stations. But these planes simply 'vanished' over the lakes.[40]

These uncanny events also happened to ships of various tonnages. The *Learfield* was a 4,453-ton freighter, and during the autumn of 1913 was travelling at speed across Lake Superior, but sank inexplicably with all hands. A few days later a grain carrier, the *James E Davidson*, crossed the same water but encountered an unusually early snowstorm and a massive sea wave, which caused considerable damage.[41] A worse fate happened to a small Canadian passenger steamer of 308 tons which was destroyed in November 1936 in Lake Huron when a single wave hit her.[42]

On 6 February 1961 a Viscount light aircraft crashed into Lake Michigan, although the weather was clear and there was no evidence of mechanical failure when the aircraft was later examined. But on the same day people had seen coloured lights flashing in the sky, which the *Chicago Tribune* said were flares from search and rescue craft. But the lights were seen all over the northeastern states.[43]

In November 1975 the North American Missile Defense Organization (Norad) despatched F-106 interceptors to check out UFOs, alleged to be seen at the time of some plane disappearances, from Selfridge Air Force base at the southern tip of Lake Huron. An object was tracked on radar over the north shore, and once appeared to shoot upwards from 26,000 to 45,000 feet, where 'it stopped awhile ... before moving up to 72,000 feet'. This event lasted six hours and was noted on air traffic control radar.[44] Nevertheless, most government reports into these 'Triangle' and 'Great Lakes' disappearances could not explain them. This point was admitted by Rear Admiral J. S. Gracey of the US Coastguard Ninth District HQ, Cleveland, Ohio, in the 1970s.

The Bermuda Triangle disappearances have been disputed, and one could argue that, as we no longer read press reports of ships and planes vanishing in the area since the 1970s, the issue was sensationally hyped up from the very beginning. In particular, a book by Lawrence David Kusche, which appeared shortly after Charles Berlitz's *The Bermuda Triangle* came out in 1975 said all the reports of disappearances were exaggerated, dubious or unverifiable. There were numerous inaccuracies and inconsistencies in witness accounts, and pertinent information that might explain the disappearances went unreported.

However, Berlitz's book contains surprisingly little about the ship and plane disappearances themselves. There were sizeable sections dealing with time warps, UFOs and other paranormal events already covered in other books and rehashed by him (and which the critics carefully ignored). Berlitz gave most of his attention to the verbatim reports of the pilots and sea captains themselves, as well as the various search crews that were sent out later, concerning their anomalous experiences at sea or in the air. Much of the book was also historical in nature and contained analysis of weather, atmospheric, geomagnetic and oceanographic oddities that go back several centuries. Berlitz himself points to other similar regions where there have been aircraft and ship disappearances, such as the south-east of Japan and the Bonin Islands, and between Iwo Jima and Marcus Island.[45] Some critics say that unexploded bombs or wartime mines in the seas went off and caused the early 1950s ships to sink. Possibly pockets of clathrate gases can reduce the density of ocean waters. Subterranean volcanic action can cause hot steam to escape, and in turn cause a massive vacuum. Oceans could be affected by different degrees of gravity, and these could affect different gradients of heat and cold in oceanic waters.

There is also the problem of 'zero magnetic deviation', and many ship disappearances may have something to do with incorrect bearings like this.[46] Magnetic and electrical forces seem to be part of the mystery, said Berlitz, especially where compass needles go awry, and where planes encounter massive air pockets and plummet like stones as the pilots fight to regain control of their aircraft. This can arise because the magnetic north on the compass is rarely the same as true north. Sailors and pilots have to make corrections that are relayed by 'isogonic lines' on navigational charts. Others have referred to 'electromagnetic vortices' that could disturb usually calm weather conditions, bringing on unexpected hurricanes and waterspouts, which rapidly overwhelmed and submerged vessels on the turbulent surface. Rough seas were known to be common in the Indian Ocean and Hawaii, and unexpected mini tidal waves have often engulfed ships. The 'seiche' – a sudden mini tsunami, possibly caused by small tremors beneath the ocean bed – have been known about for some decades, but are not usually considered to be life-threatening.

Phantom fogs

However, one 'Bermuda Triangle' report is intriguing. 'Flight 19' was a group of five planes from the naval station at Fort Lauderdale in December 1945. The crew of these five planes were flying out to Bermuda and then down in a triangular route back to base. Yet they all disappeared, and left no wreckage. What was significant was that more than one pilot reported 'white water' at the sea surface, and this has echoes of other similar reportings.[47] There also may have been a thick, white haze that even today is reported from the area. The rescue teams often reported that they had seen unexplained lights during the night of 3 October 1951 and 'dark forms and masses', but the suggestion that this was wreckage of the Brazilian warship the *Sao Paulo* that went missing in the area on that day was discounted.[48]

Strange fogs and phantom land masses have been reported throughout recent history. Massive, swirling dense clouds, which would 'strike against the houses, would break and fall down the sides in great bodies, rolling over and over', were witnessed by colonists in Connecticut in 1758.[49]

In October 1954, the coastguard buoy-tender *Smilax* was proceeding north from Brunswick, Georgia, and was heading towards the South Carolina coast. Hurricane Hazel several days earlier had destroyed navigational aids (in fact, nearly one hundred had died in North America in that month because of the storms). Soon the *Smilax* ran into a very dense fog, with the crew hearing first one foghorn and then another, although no ships were shown on the radar. Suddenly a beam of light shone down out of the fog on to the deck of the *Smilax*. There were no aircraft about at the time. It was possible, said maritime officials later, that a refraction of the ship's own headlight was seen.[50]

There were puzzling reports in June 1965 of something blocking radio transmissions from a plane in difficulty – a C-119 'flying Boxcar' – before it disappeared. The US-based Society for the Investigation of the Unexplained in the 1970s once drew up a report about a Boeing 707 on a flight from San Juan to New York on 11 April 1963. The crew noticed the ocean rising into a 'great mound' as if from an underwater explosion.[51] A pilot in 1964 from Miami who, after dropping passengers off at Nassau and returning in clear weather, noticed 'very faint glowing effects on the wings ...' while his instruments were running out of control. The glow became 'blinding' for five minutes.[52]

Another unnatural affair took place in August 1956, when the *Yamacrew*, a US coastguard cutter of the Second World War vintage, converted to cable-laying and research, was in the Sargasso Sea. In an area north and east of the Bahamas and 500 miles from Jacksonville the radar indicated, on a clear night, a huge dark land-mass ahead. Oddly, the landmass seemed to start about 2 feet out of the water, and the ship's searchlight could not penetrate more than a few feet into it. One theory was that it was some kind of sandstorm, yet the air was still. It seemed to contain an irritant of some kind, affecting the crew's breathing. Then the engine room began to lose power, and abruptly the vision vanished. There was nothing to indicate what it was.[53]

The crew on the appropriately named *Nightmare* in September 1972, which was a small diesel-powered boat, noted a 'large dark shape' blotting out the stars, and which affected compass readings. The crew also saw three moving lights in a row enter this dark area and disappear. While this was taking place, a black patch appeared in the sky above. It eventually lifted, and the lights and radio came back on.[54]

The coastguard ship *Dilligence* in November 1974, returning to Miami Beach from the Bahamas and still some 50 miles distant, noted a moving mass appearing on the radarscope, but it was somehow travelling along with the ship. Then about 10 miles away from the coast, the image vanished. Strangely, three months later, the US Coastguard *Clipper Hollyhock* had a radio failure, and could only communicate with a coastguard station in San Francisco, across the entire US continent.[55]

Fog can itself create other refracting illusions, making buildings loom closer. Leonardo da Vinci noticed that fog often filled the valley of the Po river and this made objects 'appear larger than they really are', he wrote in his *Treatise on Painting*. In Germany, the *Brocken* spectre is often shrouded in mist, and was used by Goethe in his *Faust* as the hub for witches. The *Brocken* is a ghostly figure in silhouette that looms up from the mist, its head sheathed in a coloured halo. The spectre is named after the highest peak in the Harz Mountains, created by a low sun casting the shadow of a person up on to a bank of mist or fog. The halo effect around the apparition is called a 'glory', caused by water droplets in the mist refracting the solar rays.[56] A typical *Brocken* – a tall figure complete with cowl, hood and gown – appeared with uncanny realism at a waterfall in the Yorkshire Dales in the summer of 2010. This was again created (presumably) from defracting water molecules, and a photograph of this vision was shown in the *Daily Mail* of 11 September 2010.

Similarly, the 'fogbow', sometimes known as a cloudbow or a mistbow, can be seen. Like a rainbow (already discussed in Chapter 5), it is centred on the point opposite the sun and has the same mechanism: reflection and refraction of sunlight by water droplets. But the fogbow droplets are extremely small – less than 50 micrometres across. This allows diffraction to spread the bands of colours so that they overlap and appear white. White, of course, is the total of all the colour spectrums when superimposed on each other.

Phantom subs and rockets

The steamship *Fort Salisbury* was on a sedate course through the Gulf of Guinea on 28 October 1902. During the night the crew were amazed to see a long finger of blue light stabbing into the darkness. They then saw low in the water an illuminated object, with two small orange-red lights near one end and two blue-green ones at the other. As they got closer it seemed to be a huge submarine of sorts, 100 feet in diameter and an amazing 600 feet long, at a time when submarine technology was still in its infancy. Certainly no prototype was then being built to anywhere near this length. The crew also heard strange sepulchral clanking sounds like machinery, and 'excited voices'. At the same time the object appeared to be sinking.[57]

Gradually the term 'Unidentified Submarine Objects' (USOs) gained currency. These undersea craft seemed greatly to be able to outperform conventional subs, which generally have a top speed of about 45 knots, or 50 mph. The USOs, however, were alleged to have speeds three times that, and could dive to vast depths of 27,000 feet, when most subs can reach no more than 8,000 feet.[58]

In 1963, the US aircraft carrier *Wasp* and twelve other vessels, while on manoeuvres in the North Atlantic, detected a huge underwater craft travelling at 150 knots, and diving very deep, while circling around the craft on and off for four days.[59] In July 1962 in the Gulf of Catalina, south of Los Angeles, the crew of a chartered fishing boat saw what appeared to be a large submarine in the distance, steel-grey and without markings. Some five figures were working around an odd-looking after-structure. Soon the mystery craft started to move, and gaining speed it nearly collided with the fishing vessel. It was noiseless and left no wake, and headed out to sea.[60]

At the Mediterranean port of Le Brusc late at night in August 1962, French fishermen saw a long metallic craft into which human-looking frogmen were entering one by one. The last one in waved to the fishermen before clamping the cover down. The craft then emerged from the water, and levitated for a while before bizarre red and green lights started to flash. The floating submarine began 'revolving', before accelerating away into the night sky in a blaze of primary colours.[61]

In April 1967, two Danish boys saw a 'glowing cigar' drop two long objects into the sea at Kattegat Strait some 50 miles north-east of Copenhagen. When retrieved, the objects turned out to be an unusually blended mixture of coal and lime. Other investigators using sonar detected a 20-foot long object, presumably a submarine of sorts, on the seabed, but at a depth of 300 feet it was too difficult for divers to reach. A report into the incident said that 'wheel tracks' on the seabed were detected.[62]

At night along the Adriatic coastline during 1978 there were reports of red and white lights diving into and emerging from the sea while following fishing boats. This was often accompanied by electrical interference that affected radar and radio equipment. Indeed things got so bad the fishermen refused to go to sea without naval protection. In one case in November, Nello di Valentino, the captain of a naval patrol boat along with his crew, saw a red light emerge from the sea to reach a height of more than an estimated 1,000 feet before zooming away. This apparently caused radio communications on the shoreline to be disrupted.

Many USOs have been seen in Scandinavian waters, and for some time in postwar years were assumed to be Soviet spy submarines. An object more than 50 feet long was spotted in Lake Rasvalen and other large dark objects were seen in other lakes. In November 1972 the Norwegian navy thought that a Russian spy submarine was skulking in Norwegian fjords, a controversial event that received considerable press coverage at the time. Indeed several Nato ships and helicopters were asked to help locate the mysterious intruder with the aid of sonar, and to try to make the craft rise to the surface with the aid of depth charges. At one time 'six red rockets' were seen to shoot out of the water, and further depth charges succeeded only in provoking a powerful jamming source that rendered radar and sonar pulses inoperative.[63] No mystery vessel was ever positively identified or known to have been damaged, and the Soviets vehemently denied that their navy was patrolling Nordic waters.

Many USOs have been seen off the eastern coast of South America, and in 1959 in the Nuevo Gulf, Argentina, the Argentine navy detected two mysterious submarines beneath the waves. But it was soon realized that the speed and manoeuvrability of the craft made pursuit impossible. Oddly, the unknown submersible was described as being like a huge mechanical fish, silver in colour, complete with a tail fin!

Mystery rockets have also been launched from the sea. Witnesses could see an orange glare and a rocket take off through the fog from Cocoa Beach, just off the coast from Cape Canaveral, Florida. But this was two years before a modified V2 rocket was launched on 24 July 1950.[64]

Several-radar equipped ships have complained of a similar 'ghost rocket' effect. In the summer of 1950 two policemen saw a UFO over Lake Michigan, Milwaukee. It had an eerie red glow and was visible for ten minutes. A coastguard cutter was despatched but found only a naval research vessel doing 'manouvres'.[65] In another case, the steamship *Llandovery Castle* in the Straits of Madagascar in June 1947 was bound for Cape Town, when the crew saw a brilliant light that descended to the water, and deployed a beam seemingly searching the waves.[66] The light seemed to come from a giant dark cylindrical craft, like a torpedo or rocket, with the end clipped (this common feature is often seen in UFO pictures). It was gigantic, about 1,000 feet long, and 200 feet in diameter!) The object then shot orange streamers from its rear, before taking off into space.

In November 2010 a 'mystery missile' was launched off the coast of California. It was reported that the flying object appeared to be not far from a passenger jet flying out of Los Angeles airport, but eyewitnesses said the missile was launched from the Pacific, about 35 miles from the coast. US navy spokesman Colonel Dave Lapan denied that it was launched from a US Navy ship, and added: 'Nobody within the Dept of Defense that we've reached has been able to explain what this is.' He said it was 'implausible' that a test could be carried out so near to the airport. A photo of the vapour trail published in the *Daily Mail* showed an unusually uneven launch had taken place.[67]

8 Seeing Things

How is it possible for people to misinterpret the lights and shapes that they claim to see in the night sky and in the oceans? To understand this phenomenon we must look in some detail at how the human eye perceives colours and shapes. Unfortunately, once we understand the physics of light and the biology of the eye, we might come to the conclusion that we might not trust ourselves to see anything properly!

The human eye is severely limited in what it can truly 'see'. All mammals and reptiles have sensory organs that respond to light. Visible light is that part of the EM spectrum that can penetrate the watery fluids of animal and human eyes. The retina, at the back of the eye, limits the entire range of light frequencies, and reprocesses what it can deal with. How well it does this is the key to how well we accurately perceive reality, but clearly if it cannot process the entire EM spectrum (which is technically not 'light' but 'near light') then it cannot do it well. The human eye needs several different resonator gates to tune into those 'bounced' light waves (the human ear houses something similar to a radio tuning device, which is a 'resonator gate' that tunes in to one station and not others). The human eye is sensitive only to wavelengths in the range of 390 nm to 780 nm. The nerve cells from the retina carry the picture to the brain, but they do so in a virtual digital fashion as discrete electrical impulses, not as a continuous variation of light and shade.

This implies that the world 'out there' is an ever-changing pattern of electrical impulses inside the human head. The energy levels of atoms, and hence the frequency of light that they emit and absorb, can be calculated, and thus also their chemical make-up, but this can be illusory. Perceiving the external world with human eyes is still not fully understood within the laws of human biology. There is still too much complexity in the real and paraphysical world, and in the quantum world, for the eye to comprehend.

On the whole, however, scientists think they know what is going on. When lights themselves change colour they are merely altering their frequency. We see, in effect, pulses of energy that can be measured in bandwidths, or wavelengths typical of pixels ('picture elements') in a television, and these bandwidths and pixels create the colour that we think we see. Wavelength frequencies that are too short for us to see are 'bluer than blue' – they are ultraviolet (UV) light. Similarly, wavelengths too long for us to see are 'redder than red' – they are in the infra-red (IR) spectrum. The emissions from a television remote controller are not visible to the human eye, but become so when viewed through a digital camera. This is because the remote control operates using the near infra-red frequency, and it uses a light emitting diode (LED) to transmit its signal, usually at a wavelength of 850 mm, although the silicon sensor in the digital camera will work happily as far as 1200 nm.

However, the perception of radiation frequencies is not enough *in itself* to explain why we see colours. Light constantly interferes with other EM waves as they reflect off matter. The colour of light from a hydrogen atom, for example, is characteristically different from that of a carbon atom. But the colours we see do not automatically relate to all the wavelengths of light that are reflected back into space. One of the earliest discoveries of photographers when colour film came into regular use was the fact that the colour of adjacent objects or clothing, even the blue sky, could be reflected into the image of a family snapshot, with subtle tints revealed in the final picture that were not seen by the photographer. As another example, the 'sprites' mentioned earlier as being either very red or blue are sometimes seen at different frequencies. Those sprites that are sometimes seen during electrical storms in the Colorado Rockies – which offer a clear view across the plains of Kansas and Nebraska – are often seen as orange, white or even green.[1] Another example: the brilliant blue wings of the giant Morpho butterfly does not contain any pigment, although the wings appear blue because of a complicated interference effect in the way the light photons are received.

This is a *noctilucent* or night-shining cloud, which can be seen at dawn or twilight. It is at a high altitude, reflecting the sun's rays at an unusual angle.

This is a *nacreous* or polar stratospheric cloud. Usually found in freezing temperatures it produces mother-of-pearl colours, which can become iridescent.

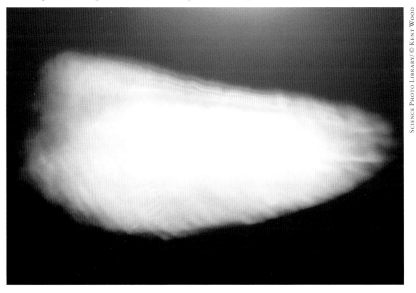

This is a *lenticular* or disc-shaped cloud. Many have been mistaken for UFOs, or even 'Cloud-UFOs', and this adds to the difficulty in determining what is actually being seen. Some seem to be exaggerated 'sundogs' or auroras and may look partly solid.

The bright part of this cloud is about to break away and form itself into a lenticular Cloud-UFO. Atmospheric pressure and temperature changes may flatten the breakaway cloud into an uncanny-looking disc.

Here we see a unique combination of sundogs, auroras and noctilucent clouds.

Another UFO-shaped cloud.

An unusual photo of small clouds taking on the spherical and angular shapes that can so often be confused with UFOs. The coloured illumination could result from unusual noctilucent effects.

A fireball landing on Earth. Astronomers often say that comets and meteorites account for most light flashes and explosions in the sky. But too often puzzled witnesses give alternative explanations to describe what they have seen.

Lightning is caused by electrical charges in the moist air that zap between clouds and the earth's surface. A negative charge at the base of the cloud makes contact with a corresponding positive charge and rebounds upwards. Many strikes can travel horizontally for up to five miles, as can be seen here.

The main lightning bolt shown here is being ejected from a menacing cloud of explosive heat and light.

Flickering balls of light contained within clouds are generally regarded as the product of lightning, although some witnesses say they are paranormal phenomena.

Charged-up lightning bolts rebounding from the ground can reach up to 300,000 amps and 1 billion volts. This can be so explosive that the originating cloud deck can be virtually vapourized into a ball of light.

Lightning bolts that appear in near-cloudless skies, as seen here, are puzzling. Some scientists believe they are caused by cosmic rays, or generated by 'plasma balls'.

A mirage of a landscape that could have been deflected forward from its original distance of ten to thirty miles.

A typical mirage, where what appear to be ancient gallions or land masses appear on the horizon. They often seem to be floating. The apparitions in this photo were actually tall buildings situated some twenty miles away from the camera.

A typical 'lake' mirage with 'trees', seen across a distance of miles. Both the lakes and the trees are actually atmospheric reflections of something else.

A view of the sea in the near distance suddenly gives way to lakes and unknown terrain on the horizon. Note the inverted objects like upside-down trees, seen in the lakes. Inversions are often observed in mirages involving the sea.

These light blobs around the Sun, which is shining in the shorter, whitish frequencies, can be caused by extremely low temperatures in northern regions, which make ice crystals bend the sunlight into curved bands.

Dr Konstantin Raudive, a Latvian psychologist, succeeded in picking up literally hundreds of unknown human voices from a bank of tape recorders left running in a sealed room, which were technically protected to block out freak 'pick-ups'.

Dr Lyall Watson, the late biologist and author of *Supernature*, also heard 'phantom voices' on blank audio tapes. He said that many scientists had picked up similar sounds in repeated experiments.

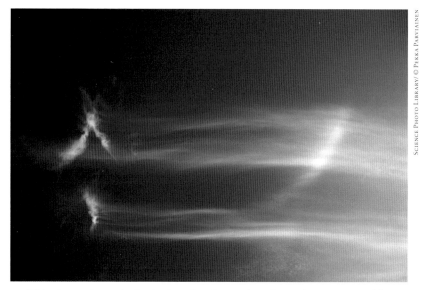

This picture shows the beginning of an inverted rainbow. These are caused by light deflections from cirrus clouds that can coast 60,000 ft up from the ground. They are known as *circumhorizon arcs*.

When the sun shines through hazy cirrus clouds it can be bent twenty-two degrees and create a 'sundog', as can be seen here.

The orange-tipped streak of light seen in the centre of this picture is either a noctilucent cloud or a sundog. It could be confused with an incoming meteorite, but the trajectory is too horizontal and the object was stationary.

These noctilucent clouds are often seen at twilight. They reflect the sun's light at an angle.

This is literally an 'unknown aerial phenomenon'. It could be the beginning of a rainbow, or frozen particles in the cloud deck could be momentarily reflecting the obscured sun's yellow light into a blur of other light frequencies.

Clouds seen at sunset can be an awesome sight, but can often result in optical illusions.

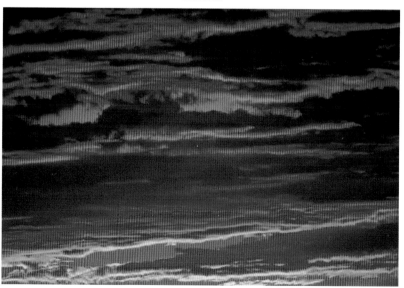

Science Photo Library/ © Stephen & Donna O'Neara

The elongated green blob could be due to squalls of dust caught up within the cloud that refract sunlight into the green spectrum. A sky that turns partly green is often a precursor to tornadoes. Alternatively, the green blob could be a Cloud-UFO, an apparition that overlaps between the atmospheric and the paranormal.

The main dark shape in the centre could be a Cloud-UFO. They are often seen against an illuminated sky and tend to blend in with it. UFOs can be confused with a variety of lenticular clouds and other natural dark gaps seen within the clouds.

Alamy Images/ © Pat Savage

Many UFOs seem to have solid characteristics, and have light clusters underneath or around their rims. The extraordinary amount of light that UFOs give off, often rotating through the colour spectrum, is as puzzling as the UFOs are mysterious.

This is a classic 'flying saucer', and possibly solid. But note the irregular dimensions and moulding of the UFO's shape, which is a common feature.

A cluster of 'Balls of Light', which could be UFOs. So often such BOLs are categorized as other types of light phenomena, such as ball lightning, 'earth lights' or perhaps space debris. The fainter second cluster of orbs, forming itself into a ring, is probably the result of lens flare.

Another 'flying saucer' with an irregular outline, looking 'apparitional'. Some UFOs, like this one, go through phases where they seem to become less solid within minutes, or change their shape.

Seawater is also a good absorber of light wavelengths, except for the shorter blue ones which are scattered effectively. The water breaks up the light in the same way that air molecules do. In other words, they take out the red and orange end of the spectrum. Rainbow colours appear in a film of oil on water when the thickness of the film is just right for optical interference to reinforce certain colours and not others, because the reflected light from the oil interferes with that reflected from the layer of water below.

There are other 'seeing' problems that have more to do with the physics of the universe than with the biochemistry of the eye. Illusions could be caused by the same mini black holes that gave rise to speculation about the mysterious event that happened at Tunguska. Many astronomers agree that the negative refraction of cosmic light around rotating black holes can change the apparent location of stars as viewed from Earth, so in theory this refraction could happen on a micro scale at the surface of the Earth.

The biology of the eye

'Contrast illusions' also cause ambiguities in the brain, due to the brain's failure to properly distinguish two objects in poor light, a very crucial factor that needs to be taken into consideration when people try to describe what they have seen 'with their own eyes'. Two squares could appear lighter or darker because the dark surround makes the square appear darker, and vice versa. A hitherto neutral grey becomes vividly contrasted and confuses the eye. Even more startling is a picture of a Rubik's Cube, with a milk-chocolate square on one side of the cube appearing distinctly orangey-pink on the other. Only when the rest of the Rubik's Cube is covered with masking paper do the two brown squares appear exactly the same. This can be an extremely puzzling phenomenon.

One explanation for optical illusions is that the retina (the main light-receiving organ of the eye) experiences eye tremor when the lens is trying to focus the image and stimulates movement in the retina.[2] Some purposely printed illusions appear to be moving, and one view is that these patterns stimulate brain regions in an area that produces sensations of movement.

Each receptor cell in the retina is wired up to a corresponding nerve cell. The retina in turn is part of the brain called the *optic tectum* upon which the visual image is mapped, much as a picture on a computer screen is mapped to individual pixels, or phosphor dots. The brain then decodes these impulses. The pigment of the object seen describes how the wavelength is absorbed, but clusters of cells known as 'cone receptors' in the optic nerve, or optic tectum, will always respond differentially with the different pigments.[3]

The eye's complexity doesn't stop there. There are *three* types of cones in the retina of the eye, each containing a different pigment that is sensitive to a given range of wavelengths of light. This means that there is a *differential* response to each colour – the pigment responsible for seeing 'blue' could not also see the 'red'. This explains how defects in the pigment receptors in some people cause 'colour blindness'. Further, these cones, and their various sensitivities, are fewer in number away from the central region. So we can only see fainter objects by averting our gaze to one side of the object. The way in which the eye adapts to light, when viewing objects at night, hints at the amount of light reaching the different parts of the retina which respond with varying degrees of efficiency. If you stare at the wall of a dimly lit room you may eventually 'perceive' tiny pinpricks of light, or perhaps ghostly swirls of lighter or darker shades. All this arises endogenously from the nerve clusters in the optic nerve.

So, despite all the complexity, the eye is *still* deficient. And this is before we talk about myopia, long-sightedness and other shortcomings in the human eye. When HD television screens were being hyped in the Western media a few years ago as being able to offer razor-sharp screen images, it was pointed out that most people's eyesight was not good enough to spot the difference between HD and the digital and analogue images already being received from their existing screens.

Another complication arises from the fact that cone receptors in the eye also get tired if they gaze at one colour for too long, and thus compensate by boosting the other colours when a different scene is viewed. For example, staring at a green-coloured picture or wall for half a minute will, when the gaze is averted, make other objects look rosier than they really are, until the light cones in the eye adjust, in the same way that the retina (which has a muscular rather than a chemical function) adjusts slowly to brighter or darker scenes. This also applies to white colours, so that they will appear to be tinted vaguely with the opposite colour that was seen earlier. Hence if a mysterious greenish or bluish light is viewed in the sky, another white object coming into view will momentarily look pinkish or orange.

Are we living in a hologram?

The problem with reality, of course, stems from the intangible ghostly nature of the atom itself, as well as from the particles and the binding energy sources *within* the atom, itself a difficult enough idea to cope with.

Quantum physics tells us that light does not flow in a direct beam, according to Dr Mark Dennis of Bristol University, but can flow like a fluid. 'It can also flow in whirls and eddies, forming lines in space valled "optical vortices" … the light all around us is filled with these dark lines, even though we can't see them', he said.[4]

But the reason light seems to flow relates to the *light particle*. The history of quantum physics pre-eminently involves this particle, known as the photon, dating back to the early 1800s when the polymath Thomas Young showed how a *beam* of light could be spread out as it passed through two narrow slits placed close together. Later of course physicists showed that it was not just a beam of light, but also the photon and other particles that could do the same thing if shot through the barrier. In actual experiments testing the two-slit theory, each time an electron or photon goes through, scientists hear a single click on the detector screen. After a thousand or so firings, classical theory stipulates that behind each slit will be a simple scatter pattern. The experiments instead show a wave pattern – just as if they had fired a stream of electron-*waves* at once!

An even greater shock came when it was, in theory, possible for a large object in the 'classical' world to behave the same way. If a bullet could be shot at the speed of light at a suitably engineered two-slit experimental wall, then the bullet itself, passing through *two* slits, would also appear as a wave. This was suggested when a larger particle, known as a buckyball consisting of sixty carbon atoms, and about a nanometre in length (invisible to the naked eye, but just about observable under an electron microscope), could be either a wave or a particle.[5]

These experiments mean that the location of the boundary between the classical and the quantum worlds remains a mystery. The Heisenberg Uncertainty Principle says that the more we know about some things in the quantum world, the less we know about others. Associated with this fact is the dilemma that when we measure a quantum system we often change it in the process, so that the observer becomes, presumably, part of what is observed. Controversially, this theory is based on the theory that the atom exists in two *potential* modes at once, and it is still not certain whether macro, solid, objects, made up of particles, could also be in two potential modes at once. Some 'anomalist' writers do take this interpretation of quantum physics seriously by suggesting that time lapses and apportations – the mysterious movement of household objects, say, from one room to another – inevitably reflect this potentiality.

Physicist Paul Davies also said that all particles everywhere are continually interacting and separating, and that everything is 'non-local'.[6] Dr Davies and many other physicists insist that quantum physics can only be understood by deploying the concept of parallel universes, with phenomena and events happening in one universe occasionally slipping into another, but only in order to explain the strange dual identity of particles. What's more, everything in space could exist in a kind of virtual world, just waiting for the uncertainty principle to become the certainty principle, after which it *then* goes into a parallel world! According to quantum field theory, empty space is actually crackling with short-lived stuff that appears and disappears in split seconds, largely, say scientists, to actually *avoid* violating the uncertainty principle.

David Bohm, prominent in postwar UK physics, did his work at the Berkeley Radiation Laboratory in the 1960s. He found that once they were part of a plasma cloud, electrons stopped behaving like individuals and started behaving as if they were part of a larger and interconnected whole. But Bohm was only repeating, in his own way, what Louis de Broglie tried to say earlier in the twentieth century. At a conference in 1927, de Broglie, one of the founding fathers of quantum theory, suggested that local electrons are somehow 'guided', or directed, by some overall 'potential'. He was severely criticized concerning this, and dropped the idea.[7] Still, most physicists accept that matter and energy are spreading ceaselessly throughout space. Even the photon 'wave packet' contains the patterns of the whole, including all of the past and all of the implications for the future.[8] If matter is wavelike, then it is composed of patterns that interfere with other patterns of energy, which could in turn explain how strange UAPs apparently deploy huge amounts of energy and often, in the process, morph from a solid to a translucent existence.

Bohm went on to conclude that the universe is a vast hologram, where every part of the image of the universe contains the whole order. Bohm's acclaimed book on the hologram theory, which came out in 1980, was entitled *Wholeness and the Implicate Order* (or the 'enfolded order'). This means that electrons are enfolded throughout the universe. Bohm and others who believe that the universe is nothing but a giant hologram say that as light waves enter the consciousness of the human brain through its sense organs, it encounters the holistic universe.

Real holograms, of course, actually exist, and have a profound bearing on how we perceive reality, and they operate by manipulating light waves. Laser light is especially good at creating interference patterns – the waves of overlapping light that particles produce at the back of the two-slit experiment. A single laser light is split into two separate beams, but these beams are highly unusual and extremely pure and 'coherent' as a light beam. One beam is bounced off the object to be photographed, and a second beam collides with the reflected light of the first. This interference is captured on film; the film looks like an interference pattern of waves itself, until another light is shone through the film, and a 3D image appears.

The hologram is thus able to reproduce the exact pattern of divergent light from an object that is photographed holistically, so that our eyes are not able to distinguish the difference between the real object and a hologram depiction of the object. The hologram seems to have an extension in space, but if you pass your hand through it there is nothing there. Similarly, no instrument will pick up any solid image or any abnormal energy. This is because the light from the object is seen from all the parts of it that are shown in the picture; in a sense the light that we couldn't see has been re-created by lasers. It is eerily convincing, and could mean in turn that we simply cannot tell whether we are living in a *hologram* world or a *real* world!

Hologram theory fell out of fashion after a while, although the theory has recently been resurrected. An article in *New Scientist* in 2009 suggested that recent experiments with a gravity-wave detector arrived at some startling results. Background noise from the universe has a kind of graininess where spacetime stops behaving like a smooth continuum, and dissolves into 'grains', just like a newspaper photograph dissolves into dots as you zoom in.

It is a frightening thought for all of us, and for scientists them-selves, if we have to conclude that all permanence, all solidity, is illusory. The fact that we can touch hard objects and feel pain still might not alter the argument because of other complicated aspects of holistic reality concerning flesh tissue, nerve endings and brain neurons, which are all built up from atoms and still subject to quantum effects. This of course remains controversial, and many scientists believe that there is a real and unbridgeable barrier between the quantum and classical world even if our maths and our theorizing says there is not. Laurence M. Krauss, a theoretical physicist at Arizona State University, feels he has to deny that quantum theory says that there is no objective reality[9] (although even a hologram is, in its own way, an 'objective reality'). But a scientist at Fermilab's Center for Particle Astrophysics, says: 'Our everyday experience might itself be a holographic projection of physical processes that take place on a distant, 2D surface.'

This is a truly unsettling conclusion. Yet it has become clear in recent years that the advance in digital and camera technology has quite markedly demonstrated how limited is the human eye when it comes to perceiving the true nature of reality. In fact, the defi-ciencies in the eye become grotesquely apparent when compared to camera technology, as we shall see in the next chapter.

9 Camera and Video Evidence

The editor of the US *Daily Herald* deplored the lack of video evidence about UFO 'sightings', and the film images that were then available. They seemed to show just 'bright lights, shaky video and fast-moving specks in the sky [that] just don't impress a generation accustomed to the movie UFOs of Hollywood'. Sam Maranto, of the Mutual UFO Network in the USA (MUFON) in reply said that no UFO picture is perfect, being more like a distortion of reality because 'anything that is genuine is emitting a frequency'. A perfect UFO picture would engender suspicions of fakery, it was suggested.[1] Many weird apparitions that had happened in the past had to be taken on trust in the absence of photographic, video or other tangible evidence.

This argument has been seriously undermined by the growth in high-quality digital cameras. The 'unseen' and the paraphysical world is now becoming apparent to us in ways it was not before. CCTV surveillance cameras have dramatically altered the debate about Fortean phenomena because they are on virtually twenty-four hours a day. Ghostly figures and inexplicable lights are being caught on tape in stately homes, pubs and even at a fire station, and many were and are posted on the internet. A mysterious glowing figure was spotted on camera gliding around the ground of Croxteth Hall in Merseyside. Blue lights were seen outside the Fire Engine pub in Bristol city centre. They merged to give the appearance of the top half of a man before drifting apart again. Still footage of this was shown in the *Sunday Express* of 25 November 2009.

A camera overlooking a courtyard at Windsor Castle some years ago recorded a thin figure wearing a centuries-old cloak and cowl after one of the windows in a sealed-off area was momentarily thrown open. Ghostly images captured on a digital camera at Tantallon Castle in North Berwick in May 2008 showed an elderly woman peering down from an ancient barred window, wearing period clothing, including a ruff around her neck. This wasn't apparent to the photographer at the time he took the photo: the anomaly was only noticed after he got home, and he sent the picture to the Edinburgh International Science Festival who were seeking such ghostly images for a study of the subject. A theory was advanced that either an unnoticed visitor had appeared at the window, or the image of the visitor had been formed by an unusual reflection of light against the wall and grille in the castle window, ignoring the crucial fact that these 'tricks of the light' were not seen by the human observer. (In other words, they were *not* 'tricks of the light'!) Furthermore, and to confound the theory that the image had been digitally altered, a woman dug out a colour film picture of exactly the same scene in a family photograph taken in 1977, showing the same image of the elderly woman at the same window. These pictures received prominent coverage in the British press.[2] They do suggest that both celluloid film and digital images can record light and colour frequencies that are not discerned by the human eye. Further, these two pictures of Tantallon Castle, separated by thirty years, tend to give credence to time-warp and paranormal theories.

A female ghost was also caught on a security camera at a Chinese factory in Perai, Penang, and shown on the internet, as was CCTV footage from the Hantu area. This latter showed two men in an office block lift, with their backs to the lift wall, with the camera lodged in the ceiling. As soon as they walked out of the lift an elderly, hunched ghost with a bowed head and bizarre shuffling gait emerged immediately from behind the men. These videos were shown on Chinese state television, with a running commentary, and were unlikely to have been faked by freelancers and shown without permission. A Japanese office block security camera showed a young woman walking through a solid wall. It too was unlikely to have been faked because the tape recorded four interior shots in split-screen fashion, making tampering more difficult (the huge quantity of tape or digital memory generated by CCTV cameras also inhibits all but the most determined fakers).

The riddle of the orbs

Lately, the rise of digital photography has given rise to another strange phenomenon – translucent plasma-like small 'orbs' that emerge in the picture and that which cannot be lightly dismissed as merely technical glitches, dust or internal light reflections.

John Pickering, an orb specialist, says the disc-like images appear in both digital and film photos, but are invariably invisible to the naked eye. They have been seen and photographed in all kinds of light settings both inside houses and in the open. They are mostly described as bluish orbs similar to gas-fire flames, but also often appear in digital photos as tinted in various shades of pink, blue and magenta.

Klaus Heinemann, an experimental physicist and space technologist, claimed there were thousands of these orb pictures available, all sent to him from the public after a report was issued on the subject with Miceal Ledwith, co-author with Heinemann. Ledwith also said he had 'more than 10,000 pictures' of orbs. 'They come in all sizes, ranging from a few inches to several feet across ... sometimes they appear alone, and at other times hundreds of them appear in colours ranging from white to blue, green, rose and even gold', he stated.

William Tiller, a theoretical physicist who spent thirty-five years researching consciousness at Stanford University in California, reminded a conference held in Sedona, Arizona, in early 2007 that circles of light that often appear mysteriously in digital photos of family and group scenes are now regarded as real phenomena in the sense of having an objective existence. Orbs can clearly be seen looming from behind another object, so the orb is part of the world outside the camera rather than a processing anomaly.[3] John Pickering says that they are not explained by sunlight, flash feedback, lens flare, raindrops and small reflective particles in the air.

Another couple of orb experimenters, Paul and Denette France, used IXUS 400 and 330 digital cameras together on a tripod to shoot the same area at an angle. They also joined a film camera to the other two.[4] Orbs have been photographed with SLR (single-lens reflex) film negatives. The pictures of orbs taken in the early 2000s showed the objects to be behind window latches and vegetation, ruling out any fakery done to the images afterwards.

William Tiller, the physicist, asserts that the orbs are also not the result of CMOS (complementary metal-oxide semiconductor) errors in digital cameras, or connected in some way with the LED functions of modern cameras. Some sceptics, however, point out that these orbs look like strange life forms, no bigger than terrestrial insects; hence they suggest the images could be extremely small bugs, flies or midges. Flying at 40 mph a bug or midge will make a foot-long red line in the picture, and could be reflecting sunlight.[5]

The theory that orbs are merely dust or pollen that the digital process has somehow captured is dismissed by those who know about digital cameras. Dust particles would appear out of focus within a very few feet, and could not possibly be seen as orbs. Raymond Buckland, a writer on esoteric matters, points out that the orbs are never in the same place in two consecutive pictures, and experiments with vapour or mist deliberately sprayed around the garden does not bring about the orbs. Expelled breath dissipates almost immediately, and would be difficult to capture on still photography or digital film. Digital cameras see further into the IR spectrum than do film cameras, and can translate these into the visual wavelength. Klaus Heinemann works with electron spectroscopy that can detect details down to the atomic level of optical resolution. He confirms that the orbs tend to show up in the more sensitive lower frequencies such as red.[6] Further, when he began examining orb pictures he found the multicoloured spheres had interior patterns resembling computer circuit boards, and each interior was unique.

Katie Hall and John Pickering, who published a book on the many pictures of orbs they have taken in their home and garden, wrote that roughly a quarter of them contained more than one orb. The authors first used a Ummax 800 pixel, but the orbs never seemed to appear in pictures taken with it. But when they bought an Olympus C200 with 2.1 megapixels and three times the optical zoom, they started to see them.[7] They also used a Pentax Optio 30 with 3.2 megapixels, and a 5.8 mm lens, on which orbs were captured. Sometimes, they said, over one hundred pictures have to be taken of outdoor scenes before they become visible. In most cases there are perfectly clear photographs in frames right next to the one with orbs in it. A relative came over with a different camera, and after a while the orbs appeared. Hall and Pickering add that no critic has ever been able to explain this anomaly in technical terms.

An image can be seen, of course, straight away from the digital picture at the back of the camera. This virtually rules out any camera fault or defect. Either the orbs are seen at once on the image, or they are not. The fact that they are not, apparently, seen on the digital viewfinder implies that the camera flash itself sparks-up this fluorescence process, and some other factor makes the orbs visible to the camera. Possibly orbs become visible when exposed to a large number of photons from a camera flash, if they are complex plasma structures. Electrons in plasma concentrations move instantaneously to a higher orbit when struck by photons from a light source. 'Plasma' and other aspects of EM fields could only become visible through the generation of large quantities of photons such as occurs when a camera flashes. This happens in microseconds which only the camera is able to record,[8] and lends credence to the reality of the orbs that cannot be seen with the naked eye (the difficulty here is that the orbs can be recorded in the open air without the need for a flash).

And when the globes of light do appear they seem to have similar, if not identical, internal detail such as circular patterns, some apparently with 'faces'. Further, the internal circular details remain the same, regardless of size, although the smaller orbs seem to be brighter than the larger ones.[9] Klaus Heinemann has said that hundreds of sequential pictures of the same orb, taken under scientifically sound conditions in rapid succession, have demonstrated that they are capable of moving up to 500 mph or more. He goes so far as to claim that orbs have a spiritual dimension, turning up in funeral pictures or weddings. He is surprised to note that if you 'ask' the orbs to appear, they actually show up – especially at happy gatherings! 'The implications for the way we view our world and physical death are enormous ...' he has stated.[10]

One photographer, Dave Juliano, on the website www.theshadowlands.net, also implied that the orbs were 'occultic', because a woman 'psychic' he was with told him to take photos in certain parts of a cemetery. Juliano had never taken an orb picture with his 35 mm (film) camera over several years of using his camera in all types of weather and lighting conditions. When told, however, by the woman to focus on certain areas, the orbs appeared in his photos.

More camera successes

Digital photography and video images make it seem, then, as if it is going to become increasingly difficult to refute the objective nature of UFOs and orbs. Image intensifiers on video cameras now exist, which means that near-ground shots and near-distance shots can be kept in focus, so that the indistinct blurring features of earlier film cameras can be kept to a minimum. This has helped in trying to assess how far away UFOs seen from Space Shuttle windows, for instance, actually are, and what approximate size they are.

There have also been conscious attempts to try to capture spook lights on film or video. Philip J. Imbrogno, an anomalist investigator and science lecturer, began an experiment in the summer of 1994 and used a film once used by astronomers called 103ao-8, now very rare, specifically to track ultra-violet light sources. The film has a sensitivity peak of around 200 mm. In the Reservoir Road area of Brewster, Connecticut, where stories of spectral lights have been common, he went out at night and thought he saw faint blobs of light moving towards him. When he developed his photos he saw a bright light source.[11] He said the light must have been in the blue end of the spectrum which could not have been captured on other types of film.

On the other hand, new camera and video technology can *refute* apparent evidence of weird events. For example, swirling and streamer lights were seen among trees and were recently photographed at Rendlesham Forest, in Suffolk, the site of a famous UFO sighting by US military personnel in December 1980, which has become a cause célèbre among ufologists. One critic said the images could be suspect because the camera was an Olympus Digital. An internet search for the Olympus software number showed a zoom lens with a low resolution, so that files would have been highly compressed and would have suffered from 'pronounced JPEG artefacts' (in effect, these would be technical limitations that could possibly create 'streamer lights'). There was also 'colour noise' owing to the small sensor and high amplification arising from the fact that the pictures were shot at 400 ISO. This could explain the streaks in the pictures.[12]

In fact, a new sensor in a digital camera aims to capture 'unparalleled' amounts of light and colour. 'Nanoengineer' Ted Sargant of the University of Toronto says most image sensors in cameras and cellphones waste most of the light that hits them. These usual sensors are known as complementary metal-oxide semiconductor (CMOS) photodiodes. Metal tracks criss-cross their surface, which carry signals from the photodiode, and block out much of the light. So just a fraction hits the pixels. Sargant said it would be better to have the sensor area above the connectors, so he introduced a proto-type-2 megapixel 'quantum film' sensor. The tracks are hidden beneath it, so the entire surface senses light. The sensing layer is a film of quantum dots – crystals of a semiconductor material just 2 mm wide. A quantum dot nanocrystal pushes electrons into a tiny region so that they no longer behave like electrons. They are excited not by other electrons, but by photons of a certain wavelength, allowing light to be sensed at that wavelength. The nanocrystal can be fine-tuned to absorb more of the visible spectrum.[13]

In any event it is not the number of megapixels that enhance the quality of colour pictures, but the sensor size. Both Fujifilm and Sony have come out with cameras that are better than the older models in capturing low-light images. With Fuji the rectangular chip at the heart of most digital cameras has been redesigned in a hexagonal shape. On this chip the tiny individual pixel sensors, called photosites, are hexagonal. This exposes more sensor surface to the incoming light. And the sensor in a Sony camera is actually a sandwich of layers with tiny lenses above some colour filters, with some wiring below, which are themselves above the actual light detectors.[14]

This means, in conclusion, that modern camera and video tech-nology, with redesigned sensors and other advances, can vastly improve the quality of any visual apparition. This implies that cameras can now see a hidden world – or at least more of the world – than we can. Even faint traces of unusual or unnatural light, some even outside the visual range, can be seen and recorded for posterity, and we have to take the word of the camera, that what it registers is true reality. Our own light sensors in our eyes, although operating in a digital fashion, do not have a layer of skin that can act as a film of quantum dots, for example.

If they did, we might be shocked by what we could actually see.

10 The Sky-zappers

The suspicion grew during the twentieth century, especially after the Second World War, that powerful radio waves or electrical impulses beamed into the sky have created enhanced auroras, fireballs and humming sounds. Inventors had a reputation in this respect. The deployment of high-powered electrical energy to bring about environmental, sound wave or seismic effects is not a new idea. In the twentieth century there were many reports of 'sky-zapping' experiments, although reports of 'sheet lightning and luminous auroral masses' were not uncommon even as early as 1821. In many cases attempts were made to divert electrical resources from the atmosphere – such as lightning – down to Earth in order to glean 'free' energy from it, and lightning rod patents flooded the archives in the nineteenth century. One of many attempts to get energy from the sky came with Luigi Galvani, who used copper masts. H. C. Vion in 1860 put up several long screens interspersed with tall metallic masts, all set on open land behind mountain ranges.[1]

In December 1882 a Professor Selim Lemstrom erected an insulated array of pointed aerials on a mountain ridge. White streamers of light resulted, which could also be heard as a 'sizzling' noise.[2] Other inventors were also busy. We learn of a 'horizontal flash of lightning followed by an aurora' that took place in 1888; and a 'spectacular aurora followed by a violent thunderstorm' was recorded in 1915.

The electrical genius Nikola Tesla (1856–1943) was the past-master at this sort of experimentation. Tesla revolutionized the world with his inventions in practical electricity, giving us the alternating current, the induction electric motor, radio telegraphy and incandescent lamps. But his most significant discovery was that electrical energy could be made to propagate through the Earth and around the Earth. He proved that EM waves in the extra long frequency (ELF) range can travel with virtually no loss of power to any point on the planet. It was rumoured that he sent electrical pulses some 20 miles without wires into the air by tapping energy with his 'Magnifying Energy Transmitter'.[3]

Tesla was mesmerized by *resonance*: small forces that can add up to create huge energies. He made fantastic claims for his Tesla coil and his other resonance-inducing pulse machines that operated at throbbing, rhythmic frequencies. Invented in 1891, the coil was really two coils inside another, so enabling two sets of currents to oscillate at the same frequency. The accumulating voltage was then hiked up higher and higher by making the secondary coil proportionally larger. This in turn could be enhanced by using the Earth's ionosphere as an electrical circuit. With a relatively modest 10 kW generator he said he could 'ping' the Earth with a rapid periodicity to allow powerful resonances to build up.[4]

In 1901 he began constructing the Wardenclyffe Tower, a giant coil 170 feet high in the shape of an octagonal pyramid, on the coast of Long Island, New York state. In one press interview he said that he could virtually bring about world-shattering earthquakes with his 'electrostatic' pulsing effects: '... I could set the Earth's crust into such a state of vibration that it would rise and fall hundreds of feet ...' he declared. He said, during the First World War, that a dozen towers, located at strategic points along the coast, could defend the USA from 'all-out aerial attack'. In 1919 the *New York Times* reported that Tesla had filed applications for a weapon that was akin to 'shooting thunderbolts' that could produce destructive effects at a distance.[5]

Continuing to promote the defence angle in the interwar years, Tesla asserted that the energy from the ionosphere could be used as a battlefield death ray.[6] He said late in July 1934, at the age of seventy-eight, that he had indeed invented yet another thunderbolt weapon capable of generating and transmitting highly concentrated particles of energy which would 'kill without trace, and could destroy thousands of enemy airplanes at a distance of 250 miles'.[7] In an article in the *New York Times* of 22 September 1940, Tesla was reported to have claimed that a 'teleforce', differing substantially from his electric power inventions, could be deployed that would operate through a beam one hundred-millionth of a square centimetre in diameter, 'and could be generated from special plant that could cost no more than two million dollars and would take only about three months to construct'.[8] This was, in effect, an adapted radio transmitter, an invisible 'beam' rather than the 'streamer' device, as he had originally described the project to financiers as a high-power radio transmitter.

But there was some truth in the vibration aspect: on more than one occasion Tesla alarmed his technicians by creating disturbing pulsating rhythms in his laboratory, situated in a four-storey building, with signals beamed at the Long Island tower. People thought they were about to die in an incipient earthquake. Locals living many miles from Tesla's Colorado Springs laboratory complained of visible arcing 'streamers' issuing from only recently erected power lines and generators. The chief investor, with growing doubts about the project, pulled out, and the tower was eventually destroyed during the First World War in case German submarines found it a useful landmark.

More electrical pioneers

Knowledge of the reflective and propagation characteristics of the ionosphere was also available to other experimenters, including Guglielmo Marconi. Marconi found that he was able to send a message across the Atlantic via the ionosphere, which bounces radio waves down to Earth. Long radio waves are reflected by the E-region of the ionosphere (about 53 miles to 93 miles above the surface), and short waves by the F-region (which is between 93 and 248 miles above the Earth). This suits television programmers, or did before digitally broadcast satellite television arrived, after which the advantages of analogue radio and microwave transmissions were greatly reduced. Much was also known by Marconi and others about the fact that water vapour along the upper boundary of the tropopause could also act as a scattering layer. (Tesla himself spoke, in 1914, of the possibility of future control of the weather via the conductive electrical flow of moisture in the air.)

Kenneth L. Corum and James F. Corum, two prewar sibling electrical researchers, had successfully produced electric fireballs in laboratory experiments using dual Tesla coils.[9] They were intrigued, early in their careers, to learn of BOLs being seen near broadcasting masts. Following Tesla's original notes, they discovered that artificial ball lightning could be created additionally by inducing an interaction between two radio frequencies. It is interesting to note that pilots who had sighted manufactured BOLs, but reported them as UFOs, obliged the scientists to discontinue their experiments because the fireballs tended to zoom towards overhead aircraft (here we have a proven link between some UFOs and airborne sightings of man-made balls of light).

Another pioneering individual was Thomas Henry Moray (1892–1972). Born in Utah, he gained a doctorate in electrical engineering in Sweden and returned to the USA, where he began research in the 1920s. Moray believed latent 'electrostatic energy' could be tapped from *seismic* ground energy, a curiously prescient understanding of the way cosmic energy forces could be linked. Even when he later realized the energy, in this instance, was coming from space and not from the ground, he enthused about the possibility that nature would soon reveal a metaphorical 'dialectic valve' that humankind could use to trap and convert energy from various natural sources.

In the 1930s he had some success with an unusual Swedish 'crystal'. It functioned on the same principle as the early 'crystal sets' which tapped radio broadcasts without the need for an external power source (but required special headphones, which may themselves have contained a power source). Nowadays it is known that other crystals and certain kinds of basaltic rock, such as galena, can trap electricity from space (the modern transistor is based on two miniaturized silicon diodes fitted back to back that are able to trap and convert radio waves).

Later on, Moray developed a device that he claimed would deliver up to 50,000 watts of power simply by connecting it to an aerial and earthing it. This device was attached to other rather obscure components (which may have been added on to confuse onlookers and patent plagiarizers), and which included standard capacitors and coils.[10] Indeed, Moray incurred much opposition, double-dealing and cloak-and-dagger conspiracies from others who either wanted to get hold of his technology or who wanted to stop its development.[11]

Developing 'beam weapons'

Both Tesla and Marconi could see the potential of the ionosphere: the former for creating 'beam weapons' and the latter to facilitate the transmission of radio waves. An actual 'death ray' was supposedly invented by one British contemporary of Tesla – Harry Grindell Matthews – just after the First World War. It so inspired public imagination that the 'ray gun' became the favoured weapon for Flash Gordon and the evil Ming – and for a generation of schoolchildren brought up on science fiction adventures. It was later employed in the sci-fi television shows *Star Trek* and *Dr Who*.

But Matthews's 'death ray' tried only to harness and use boosted UV beams of light with the hope of disabling an enemy plane's magneto. It was an early electro-magnetic-pulse device, and it actually managed to stop cars passing his house while he was experimenting at home. But it required huge amounts of power, and so further UK interest in it evaporated.[12]

But, since Tesla's and Moray's time there have been a number of attempts to modify the ionosphere, including the injection of chemical vapours, or by charging it up, or by heating it. The Second World War was a massive test arena for every type of embryonic radio and radar technology, at every kind of frequency from very low to ultra-high. It was the inspiration of Tesla, Matthews and others that spurred defence scientists on for the search for more powerful directed-energy weapons that could utilize natural radiant energy, and many of their tests resulted in LITS and arcing 'streamers'.

It was suggested that the US military needed, during the Cold War, to boost the power of directed-energy weapons as a top priority, especially as Western intelligence agencies hinted that, with a moratorium on space-based nuclear weapons in force, the USA's enemies were forging ahead. Meanwhile Stanford University physicists said they wanted to create artificial auroras borealis to better understand the physics involved. While beaming VLF radio waves into space, they found that the waves could gather extra energy from electrons housed within the Van Allen belts.[13] They had been able to magnify the power of the magnetosphere a thousand times by forcing the beams to follow the curves of the magnetic field, and then make them swing back to Earth.

Yet there remained the suspicion that beam-type experiments, conducted by civilian researchers, could ultimately have a military purpose. In the spring of 2010 residents in Cornwall had a power cut, and fishermen reported a huge blue electric flash that lit up half the sky. The source was below the horizon. Possibly, it was suggested by locals, the flash was due to frigates and warships testing weapons along the coastline.[14]

In the USA, clues came from the fact that this kind of technological experimentation to determine the power, range and the potential of beam-type weapons involved a rather large number of USAF bases leased out to, or in some way involved in, such projects. The Americans were beginning to fear that Tesla-like weapons, capable only of disabling an enemy's computers and electronics, meant that they were technologically inferior to what the Soviets were thought to be capable of. On the evening of 17 May 1974, electronic scanning instruments in the Manzano Laboratory section of Kirtland AFB in New Mexico registered a 'tremendous' burst of energy in the 250 to 275 MHz range, first noted in the upper atmosphere, and mentioned earlier in this book.[15] On 2 April 1978 eerie 'skyquakes' accompanied by streaking and blinding white lights were experienced in the tiny community of Bell Island, Newfoundland. Some described a powerful beam of light coming straight down from the sky at a 45-degree angle. One press report implied that these sky explosions were somehow connected with secret weapons tests, because the events were investigated very promptly by Los Alamos scientists.[16] Other scientists detected them via satellites, which registered a 'potential nuclear blast'.

We saw in Chapter 1 how anti-satellite test explosions using EMP weapons were done in space. But some connect the secret US development of typical pulse-based weapons with fireballs seen in the 1960s and 1970s at nuclear bases, and at Strategic Air Command bases in the USA, much closer to home at aircraft height. The Department of State claimed in June 1987 that the Soviets themselves had constructed their own pulse weapons research station near the Radio-Physics Institute near Stavropol, and that they were possibly using particle beam methods. The Department of

State pointed out that the area has the highest number of UFO sightings in the USSR.[17] It was suspected that the Russian Pamir-3U magneto-hydrodynamic (MHD) generators, using both magnetic and/or chemical-based elements, had perhaps created 'plasmas'. A danger arose from the fact that complementary US experiments were damaging US defence facilities precisely because the 'fireballs' that they caused seemed to interfere with US missile electronics (the Russians were apparently immune to this themselves as their MHD system could dispense with microelectronic components.)

Recently, a laser-guided microwave blaster was developed for the US army. It is known as the Multimode Directed Energy Armament System, and uses high-power lasers to ionize the air, creating a plasma channel that acts as a waveguide for a stream of the microwaves. The device could destroy the electronic fuse of an explosive device, or perhaps a roadside bomb. It was developed by the army's Armament Research, Development and Engineering Centre (ARDEC) in New Jersey.[18]

The HAARP project

One particular directed-energy project is known as the 'plasma lab in the sky'. This is the HAARP (High Frequency Active Auroral Research Project), and is a continuation of earlier civilian research programmes of a similar nature, such as the High Power Auroral Stimulation (HIPAS) project. HIPAS came into existence in the early 1980s, when a sizeable transmitter and antennae array moved to a site 25 miles east of Fairbanks, in Alaska. It was when Penn State University scientists won contracts from the Naval Research Laboratory in Washington DC that some concerned citizens began to worry.[19]

Several secret military research bases in New Mexico have also been experimenting in EMP or HPM technology, including Kirtland AFB, Sandia National Laboratories, the Los Alamos National Laboratory, the Phillips Laboratories and the US Defense Nuclear Agency. Many residents in the area around Kirtland complained of a continuous, low-level hum, which scientists later discovered had a vibrating frequency of 17 cycles per second with a harmonic range of up to 70 cycles per second.

The purpose of the HAARP project has been subject to conspiracy theories, despite its unclassified nature. The HAARP ionospheric experiments are conducted on a site near Alaska Route 4, about 260 miles north-east of Anchorage. Many eyewitnesses in the Anchorage area say they regularly see 'fireballs'. A series of four high mountains form part of the Wrangell–St Elias National Park. Against one of the mountains, Mt Sandford, and the Cooper River Valley, there are 48 72-foot tall metal towers, criss-crossed with guy wires and metal meshes. The total acreage makes its size comparable to six Yellowstone Parks, with a massive grid that was finally completed in 2002. The aim, it seems, is to generate a massively effective concentrated wattage power from the sky from a large ground-based utility of interconnected phased array antennae designed specifically to focus radio frequency energy, or simply to do atmospheric experiments.[20] The Ionospheric Research Instrument (IRI) – the main beam emitter – is directed by the military largely for accidental reasons in that it is based on land owned by the DoD which had earlier installed radar early-warning devices there in northern regions bordering the Arctic. Further, the project, being in Alaska, as well as being pretty close to the magnetic lines of force, is also an enterprise that needs large supplies of fuel to pump out the energy, and Alaska is well endowed with natural gas. The Arco company, the main contractor, has control over trillions of cubic feet of natural oil and gas resources.[21]

Aware of the beam-weapons controversy, HAARP's many defenders point out that even if the HAARP project could produce a powerful beam ray that would create the same devastation as would a nuclear attack, then it would only act as a 'one shot' weapon, since it would invite retaliation. Even as a weapon, the IRI project at Gakona is not in the right place; it is not directly adjacent to geomagnetic north to make full use of the power of the 'field lines' emerging from the bedrock there.

Other HAARP projects

It was soon realized that HAARP beams can definitely go into the visible range, and that more than one HAARP system was in operation. Physicist David Wilcox points to the 'unusual UFO' of massive dimensions that occurred over Norway in December 2009, and that was filmed and photographed, the images of which have appeared on several websites and in the world's press. This appeared to be a huge illuminated corkscrew spiral with internal ring-like features, with a marked fireball sequence. A blue light soared up over the north of Norway and went behind a mountain range. It seemed to halt in mid-air, then began to move in circles, and within seconds the giant spiral had covered the entire sky. This amazing thing hung in the air for nearly fifteen minutes, only to dissipate into a black hole-type shape. Finally, a turquoise beam of light shot out from its centre.

In fact, this huge coloured BOL was particularly alarming to mainstream scientists who tried to fit the apparition into a conventional astronomical, aurora or atmospheric frame of reference. Dr Wilcox says the fireball could not have been a conventional missile (i.e. one known to the world's defence analysts) because it would have to be rotating at more than 20 miles per second, and there was no sonic boom. He said also that the BOL may have been the result of a HAARP experiment that presumably the media or defence authorities did not, or could not, comment upon.[22] The fireball came from a very northerly Scandinavian direction, and other reports said that a missile, possibly a Russian Bulava ICBM, was fired from a White Sea site just east of Scandinavia.[23]

The bull's-eye type characteristics, typical of the Norway fireball, can easily appear in an ionized or partially ionized atmosphere. Tromso is at the very northern tip of Norway, close to the North Pole, where ionization causes the borealis. It is also where the spiral was most visible, directly north-west of the major HAARP installation called EISCAT – the European Incoherent Scatter Scientific Association ionosphere heating facility. EISCAT was capable of transmitting over 1 gigawatt of effective radiated power (ERP). The *India Times* pointed out that HAARP's radio waves can accelerate electrons in the atmosphere, increasing the energy of their collisions and creating a glow.[24] It even said that ramping up the energy to 3.6 megawatts – say three times that of a typical radio broadcast transmitter – can create 'artificial auroras'. Writing before the Norway spiral, the *India Times* quoted Todd Pedersen, a research physicist at the USAF research lab in Massachusetts, who said that in 2008 HAARP 'induced a strange bull's-eye pattern in the night sky [with] surprisingly irregular luminescent bands radiating out from the centre....'[25]

In the meantime, Greg Leyh, of the Stanford Linear Accelerator Center (SLAC) in California, is planning another 'sky-zapping' project that could go into the visible spectrum. He is fascinated – like Tesla – by systems that can concentrate energy 'as highly and as tightly as possible'; the art of squeezing amazing amounts of energy into a tiny spot. Leyh used two metal Tesla coils 110 feet high, hoping to launch mammoth lightning bolts across a space the size of a football stadium in the Nevada desert. The larger the secondary coil, the greater the voltage, as we have seen. He said the trick with the Tesla experiments is to energize the first coil at a relatively low voltage, say 20,000 volts: 'The secondary coil has many more turns, so when the energy is all transferred over, it can build up to millions of volts.'[26] Yet more turns and it can cause atmospheric repercussions – albeit, very, very briefly – by charging up ions that allow current to flow in a sizzling arc. He has already sent lightning bolts crackling 45 feet into the night sky.

However, *resonance*, and not only the manufacturing of electrical power, was an important factor, as Tesla demonstrated, just as the frequency of sound from an organ pipe drops as the pipe gets larger. This would make it difficult to generate lightning bolts in the sky, since the coil must create an ionized, conducting channel through which the discharge can pass. Unfortunately, artificially created ionized air soon recombines in a split second, even though the ionosphere can create powerful magnetic fields by itself. Hence lightning bolts – typical of nature – could never be man-made unless there was some way to overcome the natural tendency of the ionosphere to close up, such as manipulating plasma-type phenomena within the Van Allen radiation belts to man's advantage, and so enable the natural electrical and magnetic field of the Earth to be fully utilized.

Beam weapons in Australia?

Could other state actors, or terrorists or rogue nations, have been experimenting with beam weapons? It appears that more than one foreign power is using the vast polar regions for HAARP-type experiments. It would not be the first time that secret weapons testing in Australia had resulted in strange fireballs seen in the skies. Harry Mason, of the Western Australian event mentioned in Chapter 1, knew that something weird was brewing, and that possibly Tesla-type experiments were being used to account for the 'fireball' sightings.

Although laser beams were ruled out, there were some reports of blue-white high-voltage discharge streamers coming from domestic mains outlets as an orange beam hovered near a house. It wasn't until June 1995 that Mason suspected that perhaps terrorists using unconventional explosives might be responsible. He believed that the Aum Shinrikyo sect was somehow connected with EM beam testing from 1993 onwards after they had bought the Banjawarn sheep station in Western Australia, some 500 miles north-east of Perth, in order 'to conduct experiments there for the benefit of mankind'.[27]

Some of the weird elongated orange lights hanging vertically downwards in the sky that occurred in the 1990s, and were captured on film, seemed to have a direct focus on a 'prohibited' area of the Fremantle Garden Island Naval Base, and hence had given rise to rumours of secret weapons testing. The naval base is an island, or promontory, riddled with underground installations dating back to the Second World War, when it was used by US and Allied navies.[28]

US nuclear physicist and former US Army Lt Col. Tom Bearden has researched Soviet EM weapons and monitored Tesla fireballs for many years. Col. Bearden believes the Banjawarn incident is also an example of Soviet/Russian EM weapons technology. Other information suggested that the Aum Shinrikyo sect had connections with both Japanese seismologists and the USSR military in its very last days in 1991. In fact, officials from the Japanese government also arrived in Western Australia in the early 1990s with ideas for an 'Anti-Desertification' research project, which was received sympathetically by the Western Australian premier Richard Court and his deputy.

One prominent member of the Aum Shinrikyo sect, Hideo Murai, a nuclear physicist formerly employed at Kobe Steel Inc, was researching microwave and other EM technology. The sect had been trying to mine uranium at a ranch they had purchased in the

Australian outback – right where the alarming fireball sightings had taken place. Australia could have been chosen as the test site for such weapons because of the isolated outback, and they would be 'deniable'. The Japanese were also doing research involving the use of electronic apparatus in the Kalgoorlie, Laverton and Eastern Goldfields area. Alexander Yakovlev, senior Politburo member under the Gorbachev regime, was allegedly behind the Japanese/Russian/Aum relationship with the aim of mating Russian weapons technology with EM beam technology.[29]

The Australian Federal Police said they had evidence that Aum had samples of uranium ore shipped back to Japan. The uranium content was considered to be insufficient for bomb making, although the substance could possibly be used for radiological bombs. On the other hand, the energies involved in the Banjawarn complex were 'nuclear-bomb level', and were of 'scalar EM technology origin'.

To be sure, 'experiments' of some kind, in regard to the Eastern Goldfields events, seem to have taken place. Possibly some of the fireballs arcing up from ground level were small test missiles with electronic navigation devices, with the missiles having a payload of uranium on board to see whether this kind of arcing could be done.[30] The massive fireball that flew over Perth on 1 May 1995 was calculated to have had a trajectory that originated in Enderby Land, Antarctica, where a number of Japanese and Russian bases are situated. The fireball flew on further towards the Kamchatka Peninsula, where a huge Russian EM weapons complex/transmitter (Tx) site lies. The suspicion of a cover-up about the true significance of the Australian skyquakes arises from the fact that the police refused to get involved with any questioning about the Banjawarn LITS, despite hints that man-made events or weapons testing might have been involved, and despite knowing about Aum's uranium shipments.

11 Man-made Fireballs

Many fireballs are, almost inevitably, space debris that follows the laws of physics – what goes up must come down. The extraordinary stresses arising from zooming speeds through Earth's dense atmosphere will ultimately destroy the most solid of objects, such as space stations, made of the toughest materials.

It is no surprise that meteoritic dust grains zipping through space at ten times the speed of a bullet, or even solitary particles of solar radiation travelling at the speed of light, play havoc with lightweight space-borne machinery. Problems occur when the satellites are still attached to their massive launch rockets, but the rocket casings are made of the thinnest possible metal to reduce the launch weight. To reduce the effect of the stresses of the launch and the pressure of the atmosphere, the sides of the rocket need to flex. Hence, to prevent the entire edifice from imploding, the casing is pressurized with air, and this risks the opposite happening – an explosion. A puncture from tiny particles of debris, like flecks of paint, or micrometeoroids, risks blowing up the entire spacecraft and its fuel tank.

Indeed it is this type of hazard that microscopic pieces of debris constantly pose. It is a danger that is vastly magnified because of the enormous velocities that both the spacecraft and particles themselves achieve. In fact, the *combined* speed of the debris and the spacecraft can reach speeds that are forty times that of a bullet fired from a .38 Special. A larger object – for example, a shard of metal – would pack a kinetic wallop equivalent to more than twenty times its mass in TNT. Collisions of this type would absolutely guarantee that more space debris would be created.

The vast majority of spacecraft are lightweight satellites. And there are enormous pressures on designers to keep them that way. Satellites are really telecommunication devices rather than vehicles, consisting mainly of thin solar panels, batteries, wiring and computer components, with tiny fuel tanks and thrusters. Engineers design and build instruments sometimes to tolerances finer than a micron (a millionth of a metre), then place them on top of a rocket containing vast quantities of highly inflammable fuel. The onboard instruments are then subject to a bone-shaking ride through the Earth's atmosphere.

Despite this, some of the man-made objects crashing to Earth – like the mammoth space stations – have been massive. Weighing 75 tons, the Russian-built Skylab was eventually de-orbited in July 1979, having drifted aimlessly for five years.[1] The earlier versions of Skylab – Salyut – also suffered problems. Five of them were eventually de-orbited after only a few years in space. Salyut-7, launched on 19 August 1982, and that came down on 7 February 1991, lasted longer in space. But when it began to run low on fuel, Russian ground controllers tried to steer the 40-ton craft into the Atlantic. It was very shortly overpowered by aerodynamic effects and large chunks of it plunged into Argentinean forests, starting several fires.

Many spacecraft – including space stations and satellites – disintegrate in space as a result of component parts softening because of the heat of travelling at vast speeds, or they fold or separate. Torque pressures, convection and friction heating, vibration and pre-dynamic pressures are also involved.

Humanity's prodigious launch activities can actually be seen with the naked eye, especially at dawn or dusk. Rob Matson, an aerospace engineer with the consulting firm Science Applications International Corporation (SAIC) in California, says that 'for all intents and purposes, satellites are mirrors', and can even be seen behind cloud decks. The glinting and flaring of some satellites comes from the three flat antenna panels often made of highly polished aluminium covered with silver-coated Teflon. The sixty-six-strong constellation of Iridium LLC satellites, with their highly reflective antennae, regularly produce a flare lasting up to twenty seconds. Indeed, by astronomical standards, the Iridium flares were spectacular, reaching a magnitude of up to nine – that's somewhere between the brightness of Venus and a full moon.[2] At certain points in their orbit satellites can reflect sunlight towards the ground, forming a bright spot. One such was Echo-1, long since de-orbited, and seen by millions as a bright star, although it was barely 100 feet in diameter.[3]

The Russian space station Mir once had the largest visual magnitude, followed by a group of secret US satellites called Lacrosse, and then the Hubble space telescope. The brightest object is now the International Space Station (ISS). Some exploratory satellites, such as the Near Earth Asteroid Rendezvous (NEAR), launched in February 1996, zoomed close to Earth in January 1998 as part of a 'sling-shot' manoeuvre to gain momentum on its way to the asteroid Eros. It could be seen as a bright flash in the night sky when Nasa turned its three solar panels to inadvertently reflect the light of the sun.[4]

The drawback here is that flashing and glinting space debris is often used as a 'fall-back' explanation to account for anomalous sky activity. On 24 November 1979 at night, over a dozen sightings from Lancashire and Merseyside were attributed to UFOs, which Dr Ian Griffin, a former Nasa astronomer, said were space debris.[5] In September 2010 the former science correspondent of the BBC, Dr David Whitehouse, when asked to comment on recent episodes of people seeing sudden flashes of light in the night sky, said they could only be asteroid lights or the Iridium satellite network. He used the word 'dreaded' in reference to some witnesses' beliefs that they had seen UFOs.[6]

'Colourful acrobatics'

So although space debris probably accounts for some UAP sightings, the subject remains controversial. Specialists at RAF Fylingdales, an early warning missile centre in North Yorkshire that also backs up as a debris-watching centre, said that the 'meteor' seen over Leeds in the spring of 2007 was space debris, but data from USAF Space Command indicated no satellite decays occurred at that time. This could have been uncatalogued pieces from a Chinese payload that was destroyed a few days earlier.[7] The matter was unresolved, as usual.

Heavy metal fragments are unlike natural substances like dust or ice, and as they plummet to the ground friction can make them produce vivid colours as they break into molten blobs. As surprising as it may seem, glowing debris particles do perform some amazingly colourful acrobatic feats on their way through Earth's dense layers of atmosphere which is already loaded with polluting gases and dust. Smaller items of space debris can appear as thin streaks of light like 'shooting stars'. Even debris such as spent rocket casings that do not burn up in the atmosphere can still seem like UFOs. Over the years, amid the continuing reports of UFOs, the Russians concluded that most illuminated aerial phenomena in their own skies was the dispersal of sunlight in clouds of dust and gas formed from burning fuel and debris. The reported UFO seen in Russia on the night of 14 June 1980 was officially caused by the launch of Kosmos-1188. And the one seen on 15 May 1982 was a Meteor-2, and on 28 August of the same year, it was a Molniya-1.

One observer of a re-entering object in 1984 reported that with binoculars he could see a large spent rocket casing flaring as it fell into the ocean over Hawaii.[8] On New Year's Eve 1978 in northern Europe thousands of witnesses said they saw a trail of lights in the sky. People reported seeing the lit-up windows from a 'cigar-shaped craft'. British Ministry of Defence investigators said that this was simply space debris that had taken on a rather surrealistic appearance, and it turned out to be a returning Kosmos satellite. The metal fragments were heated by friction, forming a short chain of regularly spaced-out glowing debris. The 'windows' were hence optical illusions, as witnesses on the ground filled in in their mind's eye the outlines of a cigar-shaped spacecraft. Hundreds of witnesses across Europe described another BOL as a 'blazing rocket-shaped device', and it was captured on film. On 23 January 1983 Kosmos-1402 broke up, one part re-entering the atmosphere over the Indian Ocean, and the following month the nuclear core landed somewhere in the South Atlantic.

When witnesses reported 'an enormous star' sending pulsed shafts of light to the Earth in north-west Russia in September 1977, the Russian Academy of Sciences and the military decided to look into the matter. They later issued a report saying that this sighting was the result of a satellite Kosmos-955 launched from the Plesetsk spaceport coming down. Other bright lights seen in the night sky in

March 1993 over parts of northern Europe and Britain were alleged by NORAD to be the burnt-up Soviet military satellite, Kosmos-2238.[9]

Many satellites will lose momentum very quickly once the 100-mile altitude threshold is breached, and tend to break apart about 50 miles up. The disintegrating part travels horizontally, rather than plummeting vertically. If the event takes place within 90 miles of an observer, the debris can even appear to be rising.[10] Space debris can take in excess of thirty seconds to cross the sky – a speed that is much slower than that of incoming meteorites. The effects are most noticeable at twilight, with the sun glinting on their metallic surfaces.

Several LITS and explosions have been the result of rockets crashing into the sea, destroying expensive satellites in the process. Vividly glowing debris often rained across a huge area of the Indian Ocean and a remote part of Western Australia. The 17-ton space observation station, Nasa's Compton Gamma Ray Observatory (CGRO), came crashing back to Earth when the craft was ditched in the Pacific earlier in the year 2000. Its debris spread over 930 miles on 1 December 2001, and the fireball streaks were seen from Texas to Nebraska.[11]

Only a few months later a giant series of coloured fireballs, accompanied by thunderous explosions, streaked across the skies of the southern hemisphere. It understandably brought consternation to many East Asian citizens. What those in the Asia–Pacific region saw was the break-up of the 140-ton Russian space station *Mir*. This act of wanton space vandalism was done coolly, carefully and after much deliberation, and after decisions at the highest levels had been taken in both the USA and Russia. Indeed, as US space scientists had repeatedly pointed out, Mir was becoming dangerously decrepit, and was about to fall apart at any moment. Some forty propellant tanks, many large batteries, metal storage boxes and heavy metal bulkheads survived re-entry while glowing white and red on their way down. Then, continuing on their journey at about 37 miles up, the remainder of the structural elements started to spread out into a large area of the southern hemisphere skies. Debris came careening down over an area said to span over 3,000 miles. Mir finally erupted into a fireball over the South Pacific at 0558 GMT on Friday 23 March 2001 as the bulk of the debris – some 27 tons of it – landed in the ocean some thousand or so miles south-east of New Zealand.

'Operation Moondust'

In 1967 a secret US defence intelligence operation known as 'Operation Moondust' was set up to check into the reports of metallic objects – largely spent rockets or bits of broken-up satellites – falling out of the skies, which naturally had the Americans worried about the defence implications.

American defence agencies, as well as the CIA, soon generated large files (known as the Moondust Files (or Project)), detailing anomalous sightings of fast-descending glowing objects. Many reports were sent by US foreign embassy staff and agents to USAF analysts in the USA. What was particularly disturbing was the knowledge that very few satellites – and certainly hardly any commercial ones – were launched in the 1960s. There were, frankly, too many fireball sightings to be safely attributed to satellite burn-ups, and the term 'space debris' had not yet come into vogue.

Many of the reports were dismissed as sighting errors, or even 'flying saucers'. Growing increasingly puzzled, the Americans also scanned the foreign press for reports of strange glowing objects, many of which were indeed described as 'flying saucers'. One report appeared to refer to an object that plunged into the sea near Japan in 1960.[12] Another in the same year mentioned an item in a Stockholm newspaper that referred to a glowing object, and the Americans were alarmed when a metallic edifice fell towards the sensitive east-west German border.[13]

Objects were also falling on US territory. On 6 August 1961, in Idaho, a BOL cut off the tops of a tree and then burned its way down the trunk, leaving a burned-out hulk.[14] A red hunk of metal, 2 inches by 8 inches, landed in California on 24 August 1961. On 6 March 1963 a home in Cleveland, Ohio, had its roof damaged from a missile crashing through the bedroom ceiling, and a 7-inch thick chunk of triangular metal was found lying on the bed. But it was not a plane part, nor a rocket part.[15] More objects fell from the sky in September 1962 on to the streets of Manitowoc, in Wisconsin. They showed signs of having been machined, and of melting as a result of atmospheric burn-up, although some people attributed this to Fortean paranormal phenomena.[16]

A later series of telegrams in October 1967 sent from the US embassy in Mexico City to the USAF Field Activity Group in Fort Belvoir, Virginia, spoke of 'possible space fragments' that had caused fireball explosions in March 1965. Hundreds of other metallic objects, some spent rockets from the earlier Gemini missions, were reported on or were recovered by agents in the field.

These reports were sent in over a period of about thirty years until the Moondust files were phased out and other monitoring agencies took over, in particular the Center for Orbital and Re-Entry Debris Studies, part of the Aerospace Corporation in Los Angeles. The CIA's Directorate of Science & Technology (and its related Office of Science & Weapons Research) also undertakes analysis of space debris, thus confirming that intergovernmental suspicion about what is in space (whether UFOs or not), and who or what is sending objects into space, continues to this day. Not only do governments operate spy satellites and naturally wish to keep quiet about it, but they want to know what the other powers are doing as well. Correct orbits are listed for only two of the ten spy satellites the USA launched in 1999 and 2000. Astronomer Jonathan McDowell of the Harvard-Smithsonian Center for Astrophysics has noted discrepancies in the US debris registry, 'and it seems that the Pentagon is keeping mum as to why'. The UN Office for Outer Space Affairs confirmed this, but said it could do little about it.

The growth of 'space junk'

However, these other agencies soon became bogged down with Moondust-type reports which grew spectacularly. Space explosions and debris fireballs occurred throughout the 1970s and 1980s, and indeed increased markedly in the 1990s and 2000s. By 2010 there were more than 900 operational satellites in orbit, controlled by 115 countries.[17] According to Richard Crowther of Britain's Qineti (formerly the Defence Research Establishment) based at Farnborough, Surrey, virtually every day at least one object returns to Earth from space, ranging in size from that of a mobile phone to that of a large space station.[18]

Nicholas Johnson, head of the Orbital Debris Office at Nasa's base in Houston, Texas, says that during the 1990s about six satellite break-ups were detected each year by the Space Survey Network (SSN) programme.[19] Space junk was beginning to pile up: the number of items, by 2009, had risen by 40 per cent on just the previous four years. The US Air Force Space Command now tracks 119,000 orbiting objects that are 4 inches or more across – including 800 working satellites – and it estimates that there are half a million smaller fragments in orbit.[20]

But when do fully functioning 'satellites' become 'space debris'? Very few scientists will know for sure, because the various databases are seldom simultaneously adjusted, book-keeping style. The space debris crisis has in recent years become worrying because of the lack of openness on the part of the launching authorities, or even scientific information about when and where the debris is likely to descend.

The Russians, Chinese, Japanese and Americans have all been involved in space disasters in one way or another. Indeed, there are more spacecraft failures than the launching agencies care to admit. None of them wants to gain a reputation for creating more space debris than the others. It was already known from the early days of space rocketry in the USA and the Soviet Union from the 1960s and 1970s onwards that some 60 per cent of rocket launches experienced at least one failure, often falling back to Earth within minutes. The Kosmos series of Soviet satellites had a poor track record in this regard. Since 1962 the Russians have sent over 2,500 Kosmos satellites skyward with varying degrees of success. But in 1981 Kosmos-1275 broke up into hundreds of pieces.[21]

A major problem involves unforeseen space accidents and the unknown debris aftermath. Nowadays, most satellite launch programmes successfully complete their missions, but little can be done to prevent the jettisoned rocket boosters falling to the surface after releasing the satellite. There can be shortfalls in the launcher's thrust, so that the booster stages, dropped into the atmosphere, come down in the wrong place, burning up fiercely as fireballs in the process, causing in many cases vivid greenish-white lights and loud sonic booms, and falling on to the land surfaces when they are supposed to end up in the oceans.

The satellite Ekrain-2, which disintegrated in June 1978 as a result of a nickel-hydrogen battery malfunction, was not admitted by the Russians until February 1992. An American-launched Titan-3C Transtage broke up, and in June 1978 was detected by chance during routine Space Survey Network (SSN) tracking. On one occasion a piece of Russian debris smashed into the courtyard of a house, and toxic fuel from the rocket leaked across a large area of central Kazakhstan, starting a fire that spread over a large area.[22] A battery on board Mariner-7 to Mars, launched in March 1969, exploded in space when leaking electrolyte made the spacecraft spin out of control.[23] A few years later in the 1970s, the abandoned upper stages of seven Delta rockets exploded while still in orbit, due to the accidental mixing of propellants after the rocket had been shut down.

China's first major upper stage explosion occurred in October 1990 after the second flight of the Long March-4 vehicle. In spite of corrective measures, a later Long March launch resulted in another major explosion. This produced even larger bits of debris, with a piece of junk weighing some 2,200 lb and measuring approximately 26 feet, careening off into an erratic orbit on 11 March 2000 after being in orbit for five months.[24] Another expensive satellite launched in November 1999 by a Chinese H-2 heavy-lifting rocket inexplicably crashed into the sea.

Unfortunately some military satellites are purposely destroyed to stop them falling into the hands of foreign powers. The Soviets developed the hunter-killer series of satellites, which were tested in orbit by manoeuvring them alongside other satellites to explode and destroy their delicate solar panels and electronics. Debris from many of these destroyed satellites still remains in orbit.[25] This makes it all the more surprising that other debris was falling back to Earth within months of breaking up.

We must bear in mind that no booster or fuel module can avoid an accidental blow-up, and some rockets explode accidentally shortly after launch. Failing rockets at the launch site are not necessarily a danger to people on the ground as the sites are invariably near the coast (for example, Cape Canaveral is adjacent to the North Atlantic coastline, and Kourou in French New Guinea is adjacent to the South Atlantic). The Russians, on the other hand, using the Baikonur cosmodrome, located in the desert steppes of Kazakhstan, do not have the luxury of aborting failed launches into adjacent seas as the French, Chinese and Americans do.

Jettisoned solid and liquid fuel that can cause bright multi-coloured flares in the night sky, in addition to the disintegrating satellites themselves, can happen frequently. A local activist described in the newspaper *Segodnya* what happens when the Proton and Soyuz boosters plunge to Earth: 'They start burning in the outer atmosphere, then break up into fragments from matchbox size to fifteen metres long.' Another local official added: 'Lower down, the unburnt fuel explodes over our heads.'[26]

On occasion, sloshing liquid fuel has been blamed for rockets stalling in mid-flight or exploding. As spacecraft get bigger and electronics get smaller, up to 70 per cent of the craft could consist of propellant, so the dangers from the 'slosh' factor becomes that much greater. The upper stage rockets that had propelled the STEP-11 satellite into space in January 1994 blew up two years later. It ejected not only the remainder of its hydrazine fuel and a high-pressure tank of helium, but some 700 fresh pieces of metallic debris, which went into orbit.[27]

Large items of junk are in Geophysical Earth Orbit (GEO), some 20,000 miles up, and can take up to ten years to fall back to Earth. But those in Low Earth Orbit (LEO), at 280 miles up, can fall back within a year, and cause fireballs as they burn up. Japan's GEO weather satellite, Himawai-3, was launched in 1984, but its two upper orbit stages fell back to Earth only in 1994, accompanied by loud atmospheric explosions.

Other examples of space debris lights

Unknown date Some 270 passengers on board a Latin American Airbus A340 airline above the Pacific had a lucky escape when the wreckage of a blazing Russian satellite narrowly missed them. The pilots of the Airbus first spotted fiery debris streaking past at night, travelling at about 500 mph. Breaking up, it caused sonic booms in the Pacific areas south-west of Auckland.

1979 In August, the US embassy in Bolivia reported finding a metallic object, 'about three times the size of a basketball', covered in a metal skin, on a farm near Santa Cruz, one of two sightings of 'fireballs' in the region on the same day.[28]

1987 A dramatic Russian failure was the Polyus prototype weaponized space station, armed with laser cannons and nuclear mines. It was launched in May, but a thruster malfunction sent it plunging back to Earth, with much of its 80-ton mass surviving the fiery fall. No one was sure where it landed.[29]

1990 On 5 November, the re-entry of the Gorizont/Proton rocket body could be seen vividly across northern Europe.[30]

1993 During the nights of 30 and 31 March, bright lights were also said by astronomers and scientists at Nasa Ames Research Center, California, to be the returning Kosmos-2238 satellites. But the MoD said this incident was 'unexplained'.[31]

1994 On 28 January, the launch of Progress TM-21 from Kazakhstan prompted a great number of fireball reports.[32]

1996 The Mars-96 probe, sent up to explore the Red Planet in November 1996, was powered with some 7 ounces of *plutonium-238*. But there was an error with the fourth stage of the Proton launch rocket while it was still in Earth orbit that thrust it off course. The Russians, already handicapped by the lack of tracking ships, had only a rough idea of where it was, and blindly tried to de-orbit it into the Atlantic. In the end it burned up somewhere near Easter Island. However, the canister containing the plutonium had disappeared, and even the Americans didn't know what happened to it.[33]

1996 On 3 June, the dimethylhydrazine fuel in the upper stage of a Pegasus launch vehicle exploded two years after its launch because of excessive pressurization in the propellant tank. It produced more than 700 trackable objects.[34]

1996 In November, scientists at the Harvard Smithsonian Center for Astrophysics, Cambridge, Massachusetts, reckoned that the sightings of bright luminous rings seen in Chile were probably the aftermath of the ignition of a Russian VKS Blok-14 upper-stage rocket launched from Plesetsk.[35]

1997 A pilot reported a fireball over Chicago that streaked across the sky before going into a steep descent. It lit up a cloud deck as it passed through. NORAD believed this to be a piece of decaying rocket debris from a Step-2 Pegasus vehicle launched in 1994.[36]

1997   A cylindrical jettisoned fuel tank from a Delta-2 rocket weighing some 573 lb made a fiery landfall in Georgetown, Texas. Another object, a beachball-sized sphere made of titanium, which once contained pressurized gas used to force fuel into a Delta-2 rocket engine, also landed in Georgetown, and probably another sphere from the same rocket lay undiscovered in the Texas countryside. In the same year debris from another spacecraft was found in neighbouring Oklahoma.[37]

1997   A joint US–Japan three-stage rocket, the M-5, carrying a scientific satellite, costing £105 million and launched from a Japanese launch pad, released the satellite wrongly at a suborbital height, causing it to fall back to Earth and burn up in the atmosphere.[38] No information was released as to where the satellite and rocket landed.

1998   A Minuteman ICBM had to be destroyed in February half an hour after it had been launched, with debris spiralling out of control. This was followed some months later by a Lockheed Martin Titan-4 launcher explosion over the Atlantic in 1998, destroying a reconnaissance satellite.[39]

2000   Worldwide publicity was given to the titanium metal sphere that was seen to crash on to South African territory. Sonic booms and smoke trails were clearly heard and seen. This was just one of three large component parts of a Delta-2 launcher used to put a GPS satellite into orbit on 28 March 1996, with the other two large steel tank-like objects and conic sections of exhaust nozzles falling in rapid succession along the line of the break-up trajectory.[40]

2000   Early in the year, a Russian booster vehicle plunged into the Kazakh steppes and was not found. At about the same time, local people living on or near Easter Island in the Pacific saw a brilliant streak of light in the skies as a Proton rocket blasted skyward from a converted oil platform. This was followed by a muffled explosion and a flash of light as the rocket crashed into the sea, destroying its multimillion dollar communications satellite system.[41] The Arctic tundra around Plesetsk cosmodrome (used especially for launching military satellites into high polar orbits), situated about 500 miles north of Moscow in the largely uninhabited Arctic terrain, is littered with space debris, such as twisted chunks of Proton missile metal.

2006   On 20 May, South Africa's premier online news resource, News24.com, reported that an object crashed into the sea behind the breaker line offshore from the Port Shepstone High School. There seemed to be an explosion, although some press reports said it could have been a tornado creating a waterspout.[42]

2006   Space debris was reported from the remote Krasnoyarsk region of Siberia when objects crashed on 1 December. Interfax, a news agency service, said local residents had seen a 'flying apparatus' plunge from the sky. Multiple Russian news sources, including RIA-Novosti, Mosnew.com, Interfax and Newslab, all reported the event. It was observed by local villagers between the Siberian towns of Yeniseisk and Lesosibirsk and, according to Krasnoyarsk territory Directorate for Internal Affairs (CDIA), caused a forest fire. Specialist teams went to investigate, including Transport Prosecutors Service. Later the authorities tried to play down the incident, and there was a media clamp-down.[43]

2007   There were two satellite explosions in the first quarter of the year, involving one spacecraft and two launch vehicle upper stages which were experiencing 'catastrophic' orbital decay from highly elliptical orbits.[44] Nasa was alarmed to note that there were no fewer than five satellite failures in that year.

12 Mystery Sounds

U nidentified pulsating, humming, throbbing and hissing noises have become a worldwide phenomenon. Surprisingly, most have never been satisfactorily traced, but have been heard in all hemispheres from Hawaii to Sweden, and over many, many decades. They are often the source of complaints, and not just from people living in towns and cities.

Mysterious hums, like LITS, are another of the many profound mysteries of the Earth, and raise perplexing scientific issues. There is still confusion about whether the sounds are natural (i.e. relating to geophysical phenomena) or man-made. The dictionary defines a hum as a continuous sound that is low in tone, although this doesn't always describe what is heard. Many sufferers say the 'hum' is similar to a diesel engine, 'rumbling and pulsating'. At Victoria, British Columbia, local people said the noises sounded like the constant idling of a tractor engine. People often go for walks late at night to determine the source of the hum, but always in vain. Many of the symptoms of noise pollution resemble the flu-like sickness and headaches typical of those who experience radiation sickness. This results from a chronic, low-intensity exposure to electromagnetic radiation at radio frequencies.

It is estimated that some 25 per cent of the British population suffer from some sort of noise sensitivity. Both sound and light seemed to be connected: our sensory systems let us compress a huge range of aural and visual inputs into a much smaller range of responses, from the very quiet to the very loud. Sound waves are vibrations of the air itself, and therefore caused by tangible, real acoustic phenomena such as those produced by asteroid impacts, natural disasters, thunderstorms, weaponry, transportation systems, vehicles, planes or machinery of some sort.

Sound waves impinge upon the human ear by affecting the tiny bone at the centre of the eardrum, fancifully known as the hammer. The ear is like an analogue device, whereas the eye is much more like a digital one, responding to on/off signals. There is onward transmission to more tiny bones in the inner ear, through an opening at the bottom of a helical spiral, called the *cochlea*. The cochlea seems to be wired up to a logarithmic range, with 30,000 nerve fibres in the auditory nerve carrying all the information about the sensitivity and loudness of various sounds. Dr David Baguley, head of audiology at Addenbrookes Hospital, Cambridge, refers to an internal volume control in our ears which helps us amplify quiet sounds at times of great stress, danger or intense concentration.[1]

Another important membrane in the ear that has specific vibration sensors seems to come to terms with this, and vibrates in different places in line with certain frequencies. A man who lived in Bridgeport, Connecticut, once said he could often hear music and little voices in his eardrum. The dentist said this was due to tiny granules of silver in his tooth fillings. These granules were picking up radio signals and sending vibrations through his bones on to his face.[2]

The decibel (dB) is the unit for the loudness of a sound, or the acoustic wave pressure the sound makes. More subtle sounds are usually assessed in wavelength *cycles*. Low frequency sounds travel further, and through a greater range of materials. They can be sensed usually as a deep rumble, and can be carried over long distances. That is why we hear the bass notes of a car stereo from further away than the lighter musical tones. Sounds below the range of 56 cycles are inaudible, and said to be *infrasonic*. Infrasound waves can apparently interfere with the visual field and possibly be intriguingly connected with the EM spectrum. A curious feature of infrasound is that it can be highly directional, meaning that resonances can be set to flow in one direction only, but can also travel great distances in that one direction.

Annoying and harmful hums

The hum people hear is generally in the mid to low frequencies of around 600–100 cycles per second – and appears to rule out *tinnitus* ('ringing in the ears'), which has a higher perceived frequency. Sometimes a sustained frequency will give some clue as to its origin, but too often a variety of them are heard. In the countryside anomalous sounds of thunder and a strange mix of drones and buzzing is known as the 'hummadruz'. Local people in the 1820s heard low drones or humming noises in the suburbs of south-east Manchester. An empty office in Peter Street, Manchester, was haunted by thin piping tunes in the late 1960s. A weird plucking and strumming of a piano was heard in Humber Avenue, Coventry, Warwickshire, in the early 1970s.[3]

In Sudbury, Sussex, local people complained of hums in August 2008. Sue Brotherwood, of the local council, described it as a 'high hum'.[4] Residents in Hueytown, Alabama, heard a high-pitched noise which, according to *Fortean Times* magazine, 'resembles a dentist's drill or a fluorescent light bulb about to blow',[5] often experienced more intensely indoors than out. It cannot be blocked out with earplugs, and vibrations can even be sensed in the body. Some people are acutely aware of low frequency sounds below the 56 cycle threshold, not usually heard by others, and it can be disturbing. Possibly these sounds could bypass the ear and instead affect the brain directly.[6] For example, in 1962 an earth scientist, Allan Frey, showed that radio waves in the microwave range could be heard by some people because, he thought, areas over the temporal lobe of the brain are more than usually sensitive.[7]

Irritating noises heard in the UK go back decades. In 1995 a Scottish newspaper, the *Sunday Herald*, reported a hum that was 'first reported in the late 1950s when people in Britain began to report hearing a most unusual noise – a combination of humming, droning and a buzzing sound'. Two academics, writing in 1998, said the low frequency annoyances in the country as a whole started from the mid-1960s onwards, and reliable reports began in the early 1970s. A letter in the *New Scientist* in 1970 complained about noises heard 'only inside buildings'. In 1973 the journal reported 'fifty cases of people complaining about a low throbbing background noise that no one else can hear'.[8]

Helen Green, a fifty-nine-year-old retired postwoman, could hear continuously a low humming sound. She said it 'had a rhythm to it, going up and down, louder and softer … the vibrations were going through my body'.[9] This sensation was also perceived by teacher Katie Jacques, sixty-nine, in a suburb of Leeds in May 2009, who described the noise in similar terms: it was 'painful, vivid and constant ... it has a rhythm to it, going up and down. Like a car diesel idling in the distance. It gets worse at night.' She suggested it might be coming from a local airport, although the neighbours were apparently unaffected by it. In Helen Green's case she admitted that no one else in the household could hear it, so she dismissed it as tinnitus because the sound disappeared whenever she left the house.

Intermittent hums were also heard in Larg, on the west coast of Scotland, over the Firth of Clyde. Larg is one of several Scottish coastal towns that have been affected by the hum since about 1980 or earlier. Ms Georgie Hyslop said living in her house was like living in an 'acoustic box'. She said it 'scrambles' her radio, television and the electrical implant in her back. She said: 'As it gets worse, it progresses to an ache across the sternum, like a tight band around the chest ... my ears pop, I get stabbing pains, vertically in my head.'[10] She has been in touch with Transco, Scottish Power, the EU Environment Agency, the National Radiological Protection Board, as well as the Hunterston Nuclear power plant! (All to no avail, of course.) Dr Baguley, the audiologist mentioned earlier in this chapter, believes that most of the sounds that people complain about are more likely to have an 'environmental' source.[11] Others believe that the sounds could have a mechanical or technological origin. Some blame them on cellphone signals or masts that operate near to the frequencies people claim they hear.

Research does show, however, that much humming probably comes from mains transformers, although some unidentified sounds come from fridges, fluorescent lights or central heating and air conditioning systems. But David Deming, of the College of Geosciences in Oklahoma, points out that the hum predates cellphones in the UK. And many of the hums are heard not in large rural areas, but in small towns.[12] With power station hums, the sound is caused by vibrations from the strong magnetic field surrounding the coiled wiring needed to transform power down to manageable levels. Iron rings join up the primary and secondary coils through which magnetism pulses, and makes the iron expand and contract over 100 times a second, giving rise to 'harmonics'.

Smells and odours can arise from electrical discharges altering the energy states of various atoms, which in turn can create other chemical compounds. Ozone at both ground level and higher up in the stratosphere is created by electrical discharges like lightning. Tri-atomic oxygen (ozone) can sometimes be smelt around electrical machinery. For some time in the 1990s in Calais, France, a pervasive chocolate or vanilla odour puzzled local people, although there was no chocolate factory in the town. A mysterious sweet smell, which has descended on parts of New York city and New Jersey at least three times since 2005, returned on 5 January 2009. The cause of the syrupy scent is unknown, but it is probably caused by electrical factors.[13]

Some sounds seem to be *emulations* of other real sounds. The former television presenter, Ed Mitchell, described how his hotel room in Singapore in the early 1990s was plagued with a distant sound of drilling: 'I scoured all the corridors to find the source of the noise, but failed. The hotel staff were not aware of any improvement work or potential oil reserves...'[14]

A military origin?

More noticeable loud bangs heard in the open could also have military connections, of course. I described in Chapter 1 the number of 'explosions and skyquakes' that were confused with sonic booms, meteorite impacts, or military activity. In the USA some of them are said to come from experimental aircraft test-flown over the US south-west, such as the 'Aurora' stealth aircraft, between Nellis AFB in Nevada and Edwards AFB in California. In August 1976 the West Country in England experienced a puzzling series of loud aerial detonations. One person phoned John Dunn's Radio 2 programme to say that locals from Porlock, in Somerset, heard them each evening at around 9 p.m. Further, he said, the local community had heard them more or less regularly since 1956. In December 1976, local newspapers focused on the theory that the USA-bound Concorde was to blame.

As time passed, British meteorologists tried to explain the sounds in terms of 'air-quakes' and subterranean 'pseudoseismics'. The same theory was advanced in 1989 and 1990 when journalists referred to 'pulse-wave experiments', with seismic stations in California recording them.[15] Many sufferers living near coastal ports often blamed submarines operating at 76 Hz, and their ELF transmitters.

A house in Dyfed, Wales, in 1977 suffered from unexplained explosions and vibrations, after which 'strange glows' and fireballs were seen in the sky. Some local people said they were UFOs. The Dyfed area was the headquarters of SOSUS, or (US) Sound Surveillance System, also known as NAVFAC Brawdy, involved with underwater listening devices. There was a US-GB Nato base in the area at RAF Brawdy, and defence factors were suspected.[16]

In a similar case, a humming from outside a house near Biddenden in Kent, in October 1983, and which permeated through the walls, seemed to extend to an area at the rear of a cottage in Sandpit Woods, only a short distance away. The hums continued to January 1984, and the police were called, while electricity 'brownouts' (unaccountable dimming of lights and power) in the house led to complaints to the electricity board. A neighbouring house some quarter of a mile away also referred to humming noises. A rumour spread that the Ministry of Defence was involved, and a government laboratory or bunker-type building with blacked-out windows was found in the woods. It was suggested that trials of electronic weapons were being conducted.[17]

A strange hum was reported in Bristol in the 1970s. The *Sunday Mirror*, after an appeal, received nearly 800 letters from people saying it was a low frequency noise that sounded like an idling diesel engine. Many complained of loss of sleep.[18] Other researchers, however, said the pervasive Bristol hum was attributed to a long-range radio transmitter navigation system used by NELS – the Northwest European Loran-C System. It was suggested that transmission testing, or the repair of equipment, was being done, because the pulse width and repetition frequencies were 'similar to those used in experiments'. But Loran was being phased out and replaced with GPS, said the Radiocommunications Agency in Britain.[19]

Industrial sources of hums

Much sound comes from industrial activity, including factories, high-tension power-lines, wind turbines, water pipes and fans, and is often amplified by walls and enclosed spaces. It is known that low frequency drones and hums can travel through solid structures such as buildings in a kind of chain reaction. Sounds above 20,000 cycles are inaudible and are called *ultrasonic*. Sounds that have been attributed to industrial or electrical technology can be sensed by those living some distance away from the source. Omar Hesse and Jorge Millstein surveyed an area in north-west Argentina, and concluded that machinery was operating deep underground. Indeed, local people in the mountainous region surrounding Cachi often complained of a sensation of low rumbling or idling diesel-engine sounds.

Americans are more likely to blame strange hums on industrial or commercial activity, but too often the proof is missing. Nearly every state in the USA reports a hum, with the largest number coming from the south-west, the Pacific north-west, and south-eastern states.[20] A woman in Gaffney, South Carolina, for example, was plagued, in the mid-1960s, by eerie hums and strange mechanical sounds. She complained to the police that she believed someone was digging under her house. A thorough investigation revealed nothing. Another woman in Bellmore, New York, spoke of a buzzing and humming sound that caused her to break out in a rash. Investigators heard the sounds, but could not account for them.[21]

In an area near Taos, New Mexico, the 'Taos hum' can be heard, and was first noticed in 1991. It is often described as an engine idling in the distance, and was more prevalent during the night hours, but only intermittently. Researchers in the mid-1990s got letters from 'all over the country describing a similar phenomenon'.[22] In a similar case, oscillations appeared on measuring equipment that clearly suggested alternating electrical waves were involved, which in turn hinted at a man-made power source. Cine films also showed strange, swiftly moving lights, which appeared to plunge into the ground at the same point.[23]

Scientists found no evidence that EM signals from industrial plants caused the Taos hum, yet never doubted the nausea, dizziness, headaches and strange ear tones that these witnesses experienced. The hum also had peculiar, if not mysterious, qualities. Only about 2 per cent of the local population could hear it – about 1,400 people.[24] An inconclusive Congressional investigation was conducted in 1993–4 into the cause of these puzzling emissions.

Greg Long, active in the search group 'Northwest Mysteries', reports on strange sounds seemingly coming from underground in the south-central region of the state of Washington. 'Several large turbines' are the cause of the noise, say those citizens who have experienced it. Others asserted the sounds were like 'a loaded truck pulling up a long hill and never reaching the top'. Long suggests it might have been due to seismic activity.[25]

Hums and rumblings have also been heard in St Johnsbury, Vermont, since 1989, and in Hull and in Nahant, both in Massachusetts, and both peninsulas projecting into the Atlantic, about 7 miles east of Boston. It was louder indoors, and during the night hours, perhaps because of the masking effects of everyday urban noises.[26] An ultra-sensitive microphone could not pick up the hum. And there was no cluster of complaints that could help investigators locate its likely industrial source. As it was heard between 30 and 80 Hz where ELF signals predominate, the navy was ruled out. It remained a mystery because there were 'no acoustic signals ... or seismic events that might explain it'. In frustration, one scientific paper says the focus should be on hearers rather than on the environment.[27]

A similar frustrated complaint came from experts trying to assess the origins of the 'Kokomo hum'. An acoustic expert hired to track down the source of the hum found low frequency noise coming from a compressor and a cooling fan, but after these had been abated, the hum resumed. The expert said it was a 'non acoustic' issue. One report pointed out in exasperation that such hums were heard all over the globe. Dr Deming, the geoscientist, in fact said that the bulk of the evidence is that the hum is not, in fact, 'acoustic'.[28]

Spooky sounds from the Earth

When it comes to mysterious distant booms, rumbles and roaring sounds, the Earth itself, rather than human technology, is probably to blame. The notion that mysterious hummings and vibrations could be due to preliminary earthquake temblors (the aural equivalent of 'earthlights') goes back to the early twentieth century. A persistent hum in Hawaii once was traced to volcanism. In 1998 Japanese seismologists also noticed deep rumbles in the ground. As they did not seem seismic, it was suggested that variations in atmospheric pressure might be causing it.

In fact, an important aid to earthquake research itself comes from volcanic underground rumblings. An early generation of seis-mometers during the First World War picked up a strange, continuous hum from the northern coasts of both the USA and Europe.[29] Earthquake hums can migrate in different motions every three minutes or so probably because the surface of the crust twists in some way.[30] It is now reckoned that each latitude of the Earth has its own frequency and harmonics. 'Unnatural' radioactive signatures and microwave readings and oscillations seem increas-ingly to have an earthly origin, according to scientists at the Instituto Geofisico de Investigaciones of Argentina in June 2003.

Many geologists believe that seismic activity is often preceded by infrasonic impulses that dogs and other animals can sense. When Krakatoa erupted in 1883, it took 36,000 lives and destroyed two-thirds of the island of Java. Infrasonic waves from this disaster travelled about three times around the world. Similarly, the Lisbon earthquake of 1755 caused ripples in lakes in Norway.[31]

Some people seem to have a way of selectively tuning into 'Earth signals'. Dizziness, first experienced by victims of the 'Taos hum', also arises from seismic activity, which itself tends to disrupt local electrostatic fields, and this could be sensed by some people. Notably, that category of people known as 'sensitives', it is alleged, can predict earthquakes and, curiously, the antenna recordings of captured earth emissions can indeed be similar to the ear tones that the sensitive experience. Indeed, several women in the USA had given warnings only days before major Californian quakes. They said they experienced physical symptoms prior to quakes and erup-tions, and often heard high-pitched noises or experienced screeching ear tones. Petra Challus gave a public warning about a quake in Parkfield, California, and predicted, on 27 September 2004, a 4.8 magnitude Californian quake. The quake occurred the following day at 10.15 a.m., eleven hours after she made her prediction, although the magnitude was higher, at 6.0. She also had a 20-second left-ear tone on 28 February 2001, just two hours prior to the quake in Nisqually, Washington State, which had a 6.8 magnitude. She described it as a 'loud, highly electrical sound, much as one might imagine sound travelling through a fiber optic wire.'[32] One woman tests her premonitions at a research laboratory in Oregon. A machine is 'energized' into a special resonant state, and becomes, in a sense, a super-large antenna system stretching from southern California to Canada, and is connected between the utility power grid and the Earth's crust.

Mysterious outdoor sounds

Many hums are blamed on reflective acoustics in the outdoors. In the thirteenth century Marco Polo heard the sand 'singing' at night as he crossed the desert of Lop on his way to Cathay: 'A noise like the clatter of a great cavalcade of riders away from the road ... even by daylight men hear these spirit voices ... the strains of many instruments, especially drums, and the clash of arms.'

Richard Baxter, in a book published in 1691 entitled *The Certainty of the World of Spirits*, described how drumming, as if from a military parade, could be heard from the Drumming Well at Oundle, Northants, itself named after a drumming legend. It could surprisingly be heard from a considerable distance, and often lasted – at irregular intervals – several days. Studies later found the drumming sensation occurred when heavy rains made water levels rise rapidly, forcing out pockets of air in the surrounding underground crevices, creating a sort of gurgling noise. Caves in Dakota, Minnesota, have made loud whistling sounds from a natural opening that lets air rush in and out.[33]

A sound like a giant pipe organ has been heard infrequently in Yellowstone Park, Wyoming, for a hundred years or more, and similar sounds have often been sensed in residential areas abutting the Pascagoula River in Mississippi. At Yellowstone Lake in 1890 a vibrating sound, like phantom telegraph wires singing in the wind, 'or, more rarely, faintly heard voices answering each other overhead', was detected, according to General H. M. Chittenden in his 1915 book on the history of Yellowstone.[34] The sound was said to spookily surge louder and louder, coming closer, and then die away, lasting about half a minute each time.

Similar sounds were heard in Seskin, Ireland, in February 1928. They often sounded like 'a heavy train through a tunnel ...' In the Pyrenees, in Spain in August 1828 there was 'a dull low, moaning, Aeolian sound, which alone broke upon the deathly silence'.[35] William Corliss, in a major study of outdoor sounds mentioned in Chapter 1, lists innumerable fireball events that were accompanied by peculiar hissing or roaring sounds from the USA, England and Ireland, many in the first half of the twentieth century. He says that swishes, hums and cracklings have long been associated with auroras and meteors: 'Scientists have long regarded such observations as auditory illusions or the consequence of inexperience, mainly because no acceptable physical mechanisms existed to explain the data ...'[36]

Elementary physics can help explain why some geophysical conditions function like amplifiers and speakers, making the natural EM 'voice' of the planet louder. Consider an audio speaker. This has a cone, a diaphragm and a coil of wire, and behind the coil lies a large permanent magnet. The vibration of the sound waves, produced by the coil and transmitted through the diaphragm, is then amplified by the paper cone. A similar situation is mirrored with the Earth itself. Reflective geophysical conditions are known as 'stochastic resonance', which applies where the output is not proportional to the input.

In 2006 a team of scientists, aware of the growing concern about, and intrigued by, mysterious musical sounds coming from the earth, went to study the 'singing' crescent-shaped dunes in Morocco. They knew that sand can build up on the top of the dunes, until the grains slide down in avalanches – it is this that brings about the musical tones. The sand also appears to 'sing' on the island of Kauai, Hawaii, and also at a sand dune in Nevada. Bruno Andreotti of the University of Paris said the sound of the Moran sand dunes can be heard 6 miles away, 'and resembled a drum or a low-flying jet'.[37] He also explained how the sand bed acts like a membrane in a loudspeaker, with the grains bouncing off one another, setting up sound vibrations.

It is this amplification effect that can be emulated by phenomena from natural or man-made structures. Audible sound waves (unlike infrasounds) usually travel in all directions, like ripples, but a layer of air at a different temperature can bounce low frequency sound waves, such as explosions, down to the ground. Barely audible rumbles can be created when strong gusts of wind clash with chimneys, and the deep vibrations can penetrate even thick walls. Sound waves tend to get bent upwards if they are moving against the wind, and downwards if moving with the wind. This refraction tends to reduce noise upwind from the source of the sound, and increase it downwind. David Kemp, an auditory biophysicist at University

College, London, explains that 'when a sound travels out and up from a single source, the volume level would drop by about 84 decibels (dB) by the time you got sixty miles away. But if the sound travelling horizontally was trapped between temperature levels then the decibel reduction can be much less', although higher-pitched noises tend to pass through the inversion layer and escape upwards.[38]

In fact, the way air currents can carry sound waves over long distances can explain the existence of what many people believe are ghostly sounds. In Chicago's Woodland Lawn cemetery, an area was set aside to take the bodies of more than sixty circus people who were killed in a train crash in 1918. But many elephants and other circus animals were also killed, and visitors to the cemetery were convinced they could hear strange animal sounds on the night-time breeze. But it dawned on a police sergeant, John O'Rourke, as he walked round the cemetery one night, that the spooky circus sounds were coming from very-much-alive animals in the Brookfield Zoo a mile and a half away to the south and west. At night, when the hubbub of daytime activity ceased, the sounds from the zoo carried a great distance.[39]

Indeed, the distances that sound waves can travel can be phenomenal. In another example a massive explosion from the Buncefield oil depot in Hemel Hempstead, Hertfordshire, on 13 December 2005, caused considerable damage and killed several people. The explosion was heard more than 200 miles away in the Netherlands and in parts of southern England. This was because the cold December weather, with a blocking ridge of high pressure over the UK, made the ground colder than usual. Cold air on frosty nights close to the ground can be trapped by a warmer layer which acts as a reflective wall, and bounces sounds back to the surface, rather like a blasting car horn under a bridge. The land contours are significant, also, in regard to wind-transmitted sounds. Closer to the ground trees and buildings get in the way. In the Buncefield case the explosion was unmuffled by buildings, roads or other obstacles.

There are other telling examples of mysterious noises and their likely origins and connections with land contours. New Zealand residents said that a mysterious sound was similar to that of air being blown over a bottle top. Many feared ridicule by going public about the hum. An investigator, Tom Moir, finally recorded the hum in 2006 after he visited about thirty homes in Auckland, the largest city in New Zealand. Dr Moir says the sound, at 56 Hz, was marginally too high for it to be produced by New Zealand's 240-volt electricity home supply, which produces 50 Hz. But he discovered that the sound is most often heard in homes located in a dip in the land or valleys that could provide a shell-like effect that magnifies the sound.[40]

Strange sounds from the sea

In Britain in the 1970s people complained about a persistent humming noise in places such as Poole, in Dorset, and Hampstead Heath. A continuous hum was also heard in the Witheridge area of Devon.[41] There was a surprising unanimity in their comments: the humming was similar to a ship's engine rising and falling in volume, and the Bristol sounds were attributed by some to movements in the Irish Sea.

During the night of 21 December 1977, an unexplained aerial blast was heard all along the Cornish coast. One coastguard, used to these booms, suggested they might originate out at sea, or even come from under the sea. In the same month unexplained booms also occurred in South Carolina.[42] During 1977–78 powerful booms rocked each of the coastlines of north America.[43]

In 1979 a security officer at the Central Pier in Blackpool felt a severe juddering, which usually occurs when the tide comes in. But this time it was intermittent, and the tide was out, and it was at night. The explanation may have come from an apparition, some 50 yards out at sea, when the security man saw a large 'orange ball of light, the size of a full moon'. Over the following four hours he saw four 'star-like objects rise from the surface of the sea, go vertically upwards a few thousand feet and then accelerate away'.[44] Other witnesses testified to this event. One scientist attributed it to the same thing experienced in a quarry earlier in the year. He described it as a 'geosound', but avoided talking about the accompanying globe of light. Others who saw the light thought it could be a kind of 'earthlight' because of a large faultline some 40 miles distant from Blackpool. The Blackpool Tower, it was suggested, has aerials and transmitting dishes which could make for a 'hotspot' that would be susceptible to 'earthlight' activity.

The inhabitants of Greenland and Iceland have heard strange echoing sounds, including hums, during winter storms, as a result of the churning up of violent ocean waves travelling in different directions, as well as the vagaries of the atmospheric conditions that often predominate in the far north. Barbara Romanowicz of the University of California at Berkeley, and her colleague Junkee Rhie, collected data from networks of seismometers in California and Japan. They believed the oceans were in most cases responsible for the distant booms and rumbles, in particular the kinetic effect of the steady thumping of deep waves on the ocean floor.[45] Peter Bromirski of the Scripps Institution of Oceanography at the University of California at San Diego agreed. Ocean waves banging against the crust create both a hum and tiny earthquakes called 'microseisms', he wrote.[46] 'A huge water-filled organ pipe', according to some sources, is the acoustic result that is often detected on northern coasts of the USA and Europe.

The 'bloop'

Puzzled scientists picked up strange oceanic sounds in the summer of 1997. There was, first, a 'slowdown' sound deep beneath the waves, which occurred over the course of seven minutes, slowly dropping in pitch, rather like the sound of a plane flying past.[47] There was also a strange 'blooping' sound in the Pacific that was far louder than any aquatic mammal could have produced. It was heard repeatedly during the summer months, before disappearing for ever, it seems. These bloop sounds were traced to a remote point in the South Pacific west of the southern tip of South America, also during 1997. They were detected by the Equatorial Pacific Ocean Autonomous Hydrophone Array, run by the US navy. According to the National Oceanic and Atmospheric Administration (NOAA), the bloop rises rapidly in frequency over about one minute, 'and in sufficient amplitude to be heard on multiple sensors, at a range of over 5,000 kilometres'. Yet there were no recognized seismic events to cause this, and it certainly was not due, said the NOAA, to submarines, weapons or sea animals. Dr Christopher Fox mused that perhaps it could be due to 'ice calving'.[48]

Other studies come from the observatory in the USA called the

Array EarthScope, which is focused on the Pacific coast.[49] Scientists have suggested that when two waves from opposite directions collide but have similar frequencies, they create a special kind of pressure wave that can carry energy to the ocean bottom. This can result in a peculiar series of vibrations that are measured at a milli-hertz level which is too low to hear but still detectable with seismometers.

Further research with digital broadband technology – which has led to an increase in the number of accurate seismometers around the world – has shown that waves hitting the coast of Labrador, Canada, generate 'seismic waves in California', according to Sharon Keder, a geophysicist at the Jet Propulsion Laboratory, California. She agreed with Dr Bromirski's theory that the vibration effects were 'microseisms', and these travel at 2.5 miles per second. Other geological factors affecting noises at seabed level include storms and underwater tides crashing into steep coastlines and faultlines.[50] A scientist at the National Oceanic and Atmospheric Administration said that hydrophones installed beneath the Atlantic, Arctic and the Bering Sea are constantly producing 'mystery sounds'.[51]

Muffled booms from afar

Booming sounds have been heard in the distance, all over the globe, and to this day this remains one of the enduring features of this mysterious Earth. Ernest Van den Broeck in a scientific journal in the late nineteenth century wrote of 'mistpouffers', which were dull, distant explosive sounds heard around the coasts of northern Europe all the way to Iceland. But it soon turned out these sounds were heard, repeatedly, with varying magnitudes, all around the globe. British diplomats and surveyors heard them often. But the more scientists and explorers tried to theorize, the more bizarre was the phenomenon. As was so often the case, the direction and origin of the sounds were mysterious, apparently coming neither from the shore of lakes nor from the inland terrain nor from distant seas. One scientist was convinced, without seeming to contradict himself, that the cannon-like sounds came from the lake itself as well as from 'many miles' away, and from different directions.

In Connecticut, particularly around East Haddam, in the early nineteenth century, distant cannon-type sounds go back 'centuries'. They became embedded in Indian lore, and were known as Moodus noises,[52] and could be heard as far north as Boston, in Massachusetts, and were invariably associated with earthquake temblors. Similar loud sounds were also detected in New Hampshire from 1834 to 1846. These in particular were thought to be due to rock blasting, common in the nineteenth century with the rapid building of new towns. Detonations were also heard at intervals in Franklinville, New York, in October 1896 at intervals of five minutes. In the Rocky Mountains in the early 1800s, they were heard by explorers, and some referred to 'five or six discharges in quick succession', day or night. They said that local Indians 'were used to them'.

Similar rock blasting sounds were also heard in Talbot County in Victoria, Australia, and mentioned in June 1943 by the Australian Meteorological Monthly report, although, as is so often the case, no identified quarrying or building work was known to have taken place in the area. In the Meteorological Monthly and other science journals, an author 'asked for explanations for the occasional booming noises heard by himself and others in various parts of central and northwestern Australia ...' Possibly, it was suggested, extreme heat and cold in the Australian outback causes the violent expansion and contraction of rocks. They were also heard in British Guiana in 1868–72, and thought to come from nearby mountains.[53]

Sea and lake 'explosions', invariably muffled and distant, resembled far-off artillery, but from vague origins. They could be episodic, and sometimes heard several times a day, but then not for months or a year. G. B. Scott, during the 1870s, while on board a steamer passing through dense jungle, could hear 'dull, muffled booms of a distant cannon, sometimes a single report, at others two, three or more in succession, never near, but getting closer as time passes ...'[54] There were few firearms and certainly no cannon in the neighbourhood for miles around, he pointed out. Naturalist Alexander Humboldt and a naval crew on 20 February 1803, along an unidentified South American coastline, heard drums beating in the air. They were first thought to be 'breakers', but they soon sounded like fluid boiling and perhaps were coming from the ship itself.[55]

Gradually names began to be attributed to them. In the Ganges Delta and the Bay of Bengal they were called 'Barisal Guns', and were heard when it started to rain, and made cannon-like sounds.[56] These were sometimes attributed to surf rollers during the monsoon period at the Bay. Other explosive sounds from coasts around Europe were called 'waterguns'. In Japan, oceanic noises are known as 'uminari', and seismic events are suspected as the cause.

Off the coast of Java foghorn noises were sometimes heard. In 1825 explorers in New South Wales heard 'a gun fired at a distance of between five and six miles ... in every respect resembled the discharge of a heavy piece of ordnance ...' Some Australian sounds were like 'rumbling like the blowing off of steam from a large boiler, galloping of a herd of cattle, rumbling of thunder, blasting of mines ...'[57]

Van den Broeck's reference to muffled booms apparently coming from the North Sea was a supposition that was well known to sailors. The French referred to them as *bombes de mer*, or *canons de mer*, while in Belgium they were known as *mistbommen* (fog sounds), or *onderaardsche geruchte* (underground noises). But, again, there was no point on the horizon from which the booms could definitely be located, nor was it clear whether they were actually coming from under the waves. In Italy muffled booming sounds were *marina*, because they seemed to come from the sea. The word *brontidi* means 'like thunder', but witnesses seldom say it sounds like thunder because it is too drawn out and more muffled. Most local people attribute them to either supernatural or natural Earth-bound phenomena. Professor W. H. Hobbs, in the 1970s, said they were possibly 'seismic'.

Other strange sounds have been heard across lakes and rivers. Lough Neagh is a wide lake in Northern Ireland, across which distant booms are often heard. Seneca Lake, in New York, in 1857 was known as the 'lake gun'. In fact, 'lake roars' have become almost a sub-culture, according to William Corliss. They were the collective term for all sorts of meteorological bangs such as ice-breaks, and wind currents in underground caves close to stretches of lake water.

Some geophysical concepts were mooted that were in many ways similar to those advanced above by Scripps Institute, the University of California and the JPL. Theories about earth settling along coastal regions and river deltas were advanced. The difficulty here is that purely geological effects, even minor ones, would surely have produced waves somewhat like a mini-seiche. And although rocks do burst under seismic pressure, they would have produced a higher-pitched noise. Such rock bursts, in any event, would have occurred where daily temperature changes are prominent, such as in mountainous areas, and conceivably not in lowland areas like the Ganges Delta. The Filipinos suggested they were waves breaking on the beach or into caverns, due to changes in the weather or because of the brewing-up of typhoons. William Corliss, the geologist, mentioned the puzzling 'triplet' character of the phenomenon, when sometimes multiples of three reports were heard at one time. One British surveyor in India in the 1870s said he heard multiples of three repeatedly almost without a pause, and often with an echo!

The seasons and climate didn't seem to have much bearing on them, either. Possibly warm, calm summer days were preferred. They were picked up by barometers but not from seismographs, which is odd, and scientists thought that was significant. In Florida in early February 1895 booms were heard out at sea, and also in eastern Canada and along the New England coast. They were heard in the Cedar Keys on 28 December 1885, and in the Gulf of Mexico about 20 miles further south-east. A group of men in a sailboat about 10 miles out said the booms came at five-minute intervals, and were often heard on still mornings. Again no definite source of the sounds could be determined.

Uncanny outdoor music

The oddest features of all were the musical notes and echoes that were heard, as they were on the Palembang River in Sumatra, and the rivers of the Malay peninsula. Tuneful noises were detected complete with a cadence, and a one-two-three kind of beat in iron ships in 1870 situated off the Atlantic coast of Costa Rica and Nicaragua. Sailors thought the sounds came from fish, but critics asked why only at night and only in that area?[58]

A strange event happened while a witness was walking along the riverbed edge at Bighorn Canyon, Montana. He said he heard a howl, beginning at a high pitch and sweeping down to a bass key, and then, after a pause, reversing the tone sequence from low to high. He tried to attribute it to a 'rock-bouncing'.[59] Mysterious musical noises have also been heard at Comnrie, in Scotland, since 1597 and well into the seventeenth century.[60] In Argyllshire, in Scotland, in the mid twentieth century, trumpet notes were returned in the open air on a lower key. In Fairfax County, Virginia, musical instruments seemed to be playing in the valley areas, and were returning notes whenever a flute was played. But it was not a normal echo: often the notes were returned not in the exact scale or in the same place, or in the same octave, as if some musical rewriting had taken place. Other outdoor places where musical notes were returned include a cemetery in Paisley, in England. And in Killarney, Ireland, a bugle call was returned, but at a higher octave.

On Saddleback Mountain, in Maine, a shouted voice across the range would ricochet as a musical tone: 'A fierce Indian war whoop was returned to us in the softest musical tunes, not one of the discords being heard.' A gun was fired and the report came back in a *feu de joie* of a long musical set of chords. And slow dirge-like cathedral music was once heard along the wooded shore of Lake Superior in 1844. Along Slave River, in Alberta in Canada, the sound of engines from motorboats was returned as 'pulsating, harp-like music'.

It was this latter characteristic of mysterious sounds, the replayed musical notes in a different order, that hinted that some-thing spooky was going on. The other booming sounds, possibly having naturalistic explanations, were put into a dramatic perspec-tive. What possible physical factors could explain airborne musical tones? Weird musical events once occurred in the Rhine area in Germany when horn notes were returned 'like a swelling organ or harp music'. But heavenly music might indicate that beneficent or good-natured entities were playing tricks on people. These sounds could have been yet another example of poltergeist activity – but lurking around in the outdoors.

'Poltergeist' actually means 'noisy ghost'. In the next chapter we will look at some other outdoor events that include visions of ghostly armies, sometimes with soldiers playing instruments, but generally making a din.

13 Imprints of the Past

Harry Price, a famous psychic researcher in the early twentieth century, wrote a classic study of poltergeists, and asked the public for its experiences. He received thousands of letters. There is now an enormous literature on psychic and poltergeist activity; people of all levels of education, walks of life, and around the globe, have experienced unaccountable movements of objects that seem to defy the laws of physics, and they have heard sounds that defy the laws of acoustics. Many people also seem to have 'extrasensory perception', or have had premonitions of disaster.

The rappings on windows, sounds of furniture being moved, shuffling, queer noises, knockings on walls, are in most cases attributed to poltergeists. But of course there are many technical and scientific reasons given for them. For example, creaking and bumping sounds can be due to subsidence, or the expansion and contraction of many types of building materials and infrastructure as a result of changes in ambient temperature. G. W. Lambert (1889–1983) suggested that poltergeist phenomena, with things sliding about the room, could be due to heavy rainfall, coastal weather and other geological factors. Subterranean streams could be involved. However, this type of explanation has obvious limitations, and critics point out that no house could survive vibrations at that magnitude arising from geological factors and remain standing.[1] (Lambert's theory was partly rescued when mysterious loud noises in a house in North Yorkshire in the 1950s was found to be due to the periodic subsidence and subterranean river water being occasionally forced up a sewer at high tide.)[2]

In regard to the mysteries of light and sound, it is puzzling that people with no connection with each other have the same uncanny experience – perhaps when visiting haunted locations or seeing lights in the sky, or experiencing the same mysterious phenomena that are separated over months or years. Similarly, many people have had direct knowledge of the inexplicable that affects many others in exactly the same way. Car windscreens unaccountably being smashed while the vehicle is in transit seems to have been a regular occurrence throughout the Western world, and the usual explanations such as stones being thrown up from the road, or of missiles landing from above, have usually been dismissed. In the Home Counties in May 1952 a car travelling along a road suddenly suffered a broken windscreen. The same happened the following day to a school bus, and again between Esher and Cobham in the same month. Police said they had investigated twenty such incidents. By June the stone-throwing phenomenon had moved to Newbury in Berkshire.[3] Similar stone-throwing has also given rise to the 'phanton marksman' stories, when pellets were fired in April.[4]

There are many examples of death-dealing phantom vehicles like cars, trains and trucks. Often they are associated with tragic accidents, as in North Kensington on 15 June 1934. On that day a bus suddenly appeared out of nowhere, driving straight for a young man in a car. The man swerved and was killed when he collided with another car. The bus then simply disappeared. This apparition recurred several times over the years. On the Lamberhurst–Frant main road between Sussex and Kent numerous reports of a truck suddenly appearing, and then fading away, have caused many cars to swerve dangerously across the carriageway.[5]

Battlefield ghosts

There are numerous examples of battlefield ghosts. Groups of soldiers – even whole regiments – would mysteriously appear in period costumes. Sometimes the battles could be seen and heard as fully fledged dins, and some were silent, and sometimes the battles could be heard but not seen.[6] People also report events that have historical or geographical attributes that can be seen by one person and not another, and that could not have been known about at the time people say they see the apparition – for example, seeing only the top half of ghosts from an earlier age when the ground level was lower than it currently is, or describing ancient soldiers wearing uniforms that only a specialist historian would recognize.

For example, a contingent of Roman soldiers was heard and seen in York at the Treasurer's House, as it was then, in 1953. A young plumber, Harry Martindale, was working in the cellar when he saw a legion of men march through a solid wall and disappear at the far side of the building. The property was on the site of a Roman road at a slightly lower level, so the man's view of the soldiers was of them with the lower extremities of their legs hidden below the floor, including those of the horses.[7] However, Mr Martindale's description of the ghostly Roman soldiers, in a television programme on the paranormal broadcast in the 1980s, stunned even historians who hitherto had to interpret the way Roman soldiers looked, and the apparel they wore, from a motley collection of ancient manuscripts and wall paintings. He said, 'The first thing I looked at was the helmet. It came straight underneath the chin. From the bottom of the helmet I could see there was a growth of hair; most of them had beards. On the top half [of the body] they wore broad bands of leather joined together to form a jerkin. On the bottom half they wore a skirt.' He said the soldiers all carried a short sword, and one soldier had a large round shield with a raised boss in the middle.[8]

And in Lulworth, Dorset, in 1935, the indistinct forms of phantom Roman armies could be seen on foggy nights, and sometimes prehistoric warriors as well. Witnesses spoke of the thump of tramping horses and men. In 1936 a writer spoke of 'an army of skin-clad folk'.[9] Several people saw a phantom army in November 1960 on a road near Otterburn, Northumberland, the site of a fourteenth-century battle.[10]

Phantom regiments are treated as a portent of battles to come, perhaps, or mirages of distant battles, or a function of second sight, or a meteorological freak. Occultic literature is filled with the sights and sounds of murders and mass killings, arising from unusually powerful emotional events that occurred in history and are somehow 'replayed' at the site of battlefields, war zones and military barracks.

A writer on the occult, T.C. Lethbridge (1901–71), once said that local traumatic events, even in the open, could result in the 'bank of depression' that he and his wife had experienced on a beach in Devon which had 'somehow imprinted itself on the atmosphere'. This bank was clearly defined; it was possible to literally step in and out of it, a kind of depressive invisible wall. Could this be due, he asked, to subterranean water, or ocean water, creating an electrical field that could be sensed by some people?[11]

Two women who visited Dieppe in France in 1951 claim they heard a virtual re-enactment of an allied assault against the Germans at the village of Puys which had taken place on 19 August 1944.

Visions, spectral voices and sounds are also to be found in regional areas, such as South and East Asia, the Middle East and Russia, where there have been widespread pogroms, sectarian massacres and natural disasters, but the ghostly aftermath is often obscured from Western researchers or misunderstood. Phantom soldiers in ancient Hindu costumes have been seen in India.[12] Ghostly images have been noted in Hawaiian and Assyrian texts. There have been periodic accounts of Cambodian ghosts, victims of the 'Killing Fields' massacres of the 1970s. There were spasmodic reports of wailing ghosts who were victims of the 2004 Boxing Day tsunami that devastated parts of Indonesia.

Raymond Buckland, a US specialist and author on the occult, listed many US battlefield ghost sites: Antietam, Maryland; Chickamauga, Tennessee; the Alamo, San Antonio, Texas. He reminds us that at the battlefield at Gettysburg nearly 8,000 soldiers died during that conflict in July 1863. At the Emmitsburg Road site, people could hear cannon, and soldiers running on hard ground, even though the area was covered with dense brush.[13]

In the early twentieth century an anthropologist and religious scholar, W.Y. Evans-Wentz, toured the Celtic regions of Britain in search of fairy-type stories. He published a large volume about apparitions and spectral conflicts on moonlit meadows that were seen by local people to be thronged with men battling in medieval armour. Wentz concluded that these were an 'afterimage' from nature, and perhaps had something to do with the atmosphere 'upon which all human and physical actions or phenomena are photographed or impressed'.[14]

Ghosts can materialize when people are awake or dozing, during the day and at night. They can be seen to vanish through walls or disappear in a flash, and can appear as solid, realistic individuals, or sometimes are merely translucent. RAF Mildenhall is in remote Suffolk Breckland, and a ghost known as 'Old Roger', dressed in a fluttering cloak and with long hair streaming in the breeze, would play an ancient flute. It is claimed he could 'whistle up' a sandstorm, and did this twice during the last war to protect the airbase from German raiders. This was apparently true, because when the wreckage of a German plane was found a few miles away on the east coast, it was said to be coated with sand that had gone into the plane's mechanism 'at every point'. Weathermen, who heard of this rumour, said it was impossible for sandstorms of that intensity to arise anywhere in England.[15]

But what is discernibly different from ordinary solitary ghosts is the vastness of the apparitions, and the length of time they take before they disappear, and also the loud battle sounds and number of people in the open air who have seen the apparitions, and who even call upon others to witness the bizarre events. For example, Sir Edmund Verney, standard bearer to Charles I, was killed at the Battle of Edgehill, Warwickshire, on 23 October 1642, the first major encounter of the Civil War. Witnesses at Claydon House, the home of the Verney family, south of Middle Claydon, Warwickshire, say they see apparitions of the re-enactment of the battle by phantom armies on every anniversary of Sir Edmund's death.[16] A pamphlet published shortly afterwards spoke of the re-enactment of this battle around the following Christmas, although the Civil War had moved out of the area. Further, the ghostly sounds and sights of the battle, including the sight of ensign flags, drumming and 'musquets' going off, was heard after midnight. This was witnessed by many country people, such as shepherds, at the time, and the vision was repeated on several nights. When King Charles heard of these reports he sent officers to take sworn statements, and the King was said to have authorized the distribution of the pamphlet. Some recent writers have also recorded accounts of the ghostly apparition that sometimes recurs on the anniversary of the battle.[17]

An aurora was said to have occurred over Northumberland, at the height of the Civil War, when an armed rebellion led by the Earl of Derwentwater took place. The earl was captured and taken to London and executed for high treason. As his hearse was driven back up north, sheets of luminous colour appeared in the skies as a kind of omen. But other witnesses say phantom armies could be seen over London and Oxford, and people saw 'swords drawn, and Armies fighting in the Air'. The sky battles were also seen in Worcestershire, with armies with weapons, plus the smell of gunpowder and sulphur. However, Paul Simons, a weather writer, discounts these visions, saying sparked-up EM fields can appear like this to some people, quoting an eyewitness at the time who did think it was some kind of aurora display, with movements within the sheets of colour 'disposing itself into columns or pillars of flame ... and after many undulatory motions and vibrations, there appeared to be continual fulgurations, interspersed with green, red, blue and yellow'.[18] It is difficult to understand, however, how several large groups of people, in cities separated from one another by over 100 miles, could confuse ghostly and spectral events with a natural outdoor phenomenon that was a separate event, especially as the aurora itself was clearly identified by some and not others.

As an example of how specifically detailed witnesses could be when it comes to describing what they see, take the 'ghost' regiment that was seen east of Blencathra, Cumberland, in mid June 1735 and again two years later. This 'vast army' suddenly appeared on horseback, five abreast, following along behind a straggling group of horsemen returning from a hunting expedition. These horsemen had no idea which militia it was or who it belonged to. The phantom regiment existed for several hours until the light faded. There was no war in the area until ten years later. Then, again, the apparition was seen by a group of twenty-six people, but before the Scottish wars started, and the wars were not in that actual region. What is significant is that witnesses with no knowledge of military matters described manoeuvres in detail: '... he frequently observed that some one of the five would quit rank, and seem to stand in fronting posture, as if he was observing and regulating the order of their march, or taking account of the numbers, and after some time appear'd to return full gallop to the station he had left ...' Sceptics, writing in 1821, who denied militia manoeuvres were taking place there, said it was an 'optical illusion'.[19]

Another phantom army was seen in December 1678 along the Purbeck Hills, in Dorset. A local military squire, Captain John

Lawrence, and a small group of workmen, could see 'a vast number of armed men, several thousands, marching from Flowers Barrow, over Grange Hill'. This was a distance of at least 5 miles. As the army came closer they could hear 'a great noise and clashing of arms'. They managed to round up witnesses to this spectacle, some 100 or so. In fact, a defending militia was hurriedly deployed, and Lawrence himself rode to London to warn of a 'possible Popish uprising'. But no attack came, and the vision disappeared.[20]

The Battle of Naseby, Northamptonshire, was fought on 14 June 1645 between Royal forces and Cromwell's men. For years afterwards there was a re-enactment of this battle, accompanied by battle sounds. Sometimes, as in the Earl of Derwentwater case, these took on weird cloud or paranormal shapes.[21]

A phantom army of some twenty-six men was seen on 23 June 1744 on the mountain of Souter Fell in Cumberland, in the Lake District. Troops were seen riding up and over the mountain. People looking out from their cottage windows recalled the event. This was not a mirage of an event on the other side of the mountain or further away, as was suggested, and there was no road on the far side, according to anomalist writers Bob Rickard and John Michell. They also relate in their book *The Rough Guide to Unexplained Phenomena* that in 1785 soldiers were seen marching through the sky in a town in Silesia, in Germany, at the time of the funeral of General von Cosel. Some twenty witnesses saw yet another 'army' marching through the sky on 3 May 1848 at Vienne, France.[22]

A large phantom army was seen in the village of Buderich, Westphalia, on 22 January 1854. Platoons of infantry, cavalry and wagons were all distinct in appearance, and moving across open country. There was even the glinting of 'firelocks' in the fading light, and the colour of the cavalry uniforms was noted. The army was advancing towards a thicket of the wood of Schafhauser. There were over fifty witnesses to this remarkable scene. Yet there was no battle going on at the time: they could have been troop exercises, of course, although the vision appeared late in the day just before sunset, and any exercises would have been known about by local people.[23]

Wroxham Broad in Norfolk has long had legends of marching Roman soldiers. In this case they are invariably seen striding over the Broad where a lake should have been, so that both the soldiers and the terrain itself became part of the time-slip. Although archives show the Norfolk Broads were actually flooded marshland in many places during the Roman era, one theory is that the Wroxham Broad itself was actually dry land near a Roman amphitheatre, and was later purposefully flooded.

A man called Benjamin Curtiss, writing in 1603, described how he and his friends were swimming across the Wroxham Broad when the lake slowly drained away and the young men found themselves in the middle of this amphitheatre, and they themselves were dressed as Roman soldiers! Over 100 years later, in 1709, the Revd Thomas Josiah Penston, in *The Gentleman's Gazette*, recorded how he and friends were picnicking on the banks of the Wroxham lake when a 'golden chariot' drawn by 'ten white prancing stallions, about a dozen lions led in chains ...' passed nearby, before vanishing beside the lake. This vision takes some beating for the sheer number of characters and animals involved in a massive carnival-type parade, which would only be computer generated if Hollywood tried to emulate it. As well as the lions and stallions there were 'several hundred' long-haired slaves, or 'sea-soldiers, all chained together', accompanied by countless noisy drummers and buglers, plus 700 or 800 hundred horsemen in a long procession of 'archers, pike-men and ballistic machines'.[24]

A soldier, Alex Ainscough of Wallasey, Cheshire, related a spooky time-warp experience during the retreat of 1918 in northern France when he seemed, as with the swimmers in the Norfolk Broads, to take on the identity of ghostly entities them-selves; to be part of the action, so to speak. The soldier in this instance was billeted in the suburbs of Rouen, a town he knew nothing of. During a prolonged lull before he had to return to duty, he took a stroll down some of the narrow side streets. A strange chill struck him as he seemed to suddenly become familiar with his surroundings: 'I was marching with seven other men at the head of a column. We were all clad in black chainmail, and we were all tall. In front rode three men on horseback, also clad in black mail. We were going to see the burning of Joan of Arc.' Suddenly the vision vanished, and the soldier soon found himself outside some kind of market, and on the pavement were marks cut into the stones stating that Joan of Arc had been burned on the spot.[25]

Phantom planes

An aircraft seen in 1934 in the skies over the Wirral peninsula of Merseyside appeared to be in serious trouble. A bus driver and his conductor near the town of Hoylake on the coast pulled up abruptly to watch in horror as the plane dived into the sea to sink beneath the waves. A few months later a woman saw another plane crashing at the very same spot.[26] An RAF search and rescue attempt for both planes revealed nothing. For the first time the term 'ghost aircraft' was used in the media – in other words, the planes were not real, and were similar to the apparitions that psychics at the time understood were connected with violent death and tragedy. The carnage of two world wars is an obvious factor in ghostly pilots and haunted hangars and airfields. A writer for *Prediction* magazine in June 1942 said that this could have been an 'etheric reflection' of earlier air disasters.

Indeed, the sighting of phantom planes gained momentum during the Second World War. In 1941 a man returning to a hotel in the Thames Valley heard what he thought was an English plane, with searchlights sweeping around, but there was no alert. He saw the plane attempt to land in some nearby fields; it flew over his head, out of control, 'whirring and spluttering'. When the inevitable crash occurred, the sound of it was 'curiously hollow-sounding and reverberating'. He ran to the site but to his astonishment found nothing – 'it had vanished abruptly and inexplicably'.[27] The witness to this ghostly event, a Mr Elliott O'Donnell, made enquiries and found that two pilots of an unidentified biplane had crashed on that spot ten years earlier.

In 1942 someone heard what he thought was a German Messerschmidt ME62 crash in a field close to an aerodrome in the South East, although, nothing was found. And later in the summer a badly shot-up Lancaster bomber – this time a real one – was struggling to return to base after a raid on Germany. Suddenly the air gunner, through dense fog, saw a light flashing ahead, spelling out a direction path in morse identification letters that were presumably coming from the flight commander. The flight crew followed the light to the airfield, and landed safely. But they shortly afterwards found out that the flight commander had been shot down in flames the day before.[28]

And bizarrely, in November of that year, RAF Fairlop in Essex was put on alert when an apparently unidentified plane approached, assumed to be a German raider. A dark shape loomed in the sky and searchlights were switched on. It turned out to be a Sopwith Camel First World War biplane, but abruptly disappeared. But this was not before an alert pilot had noted that the phantom plane included a relocated cockpit, petrol tank and other changes. The leading pilot known to have flown this modified little fighter was one Lt George Craig of 44 Squadron, who was killed in action.[29]

Some years ago I recall reading an autobiography of a Second World War airman which related a tale of a spooky voice that was heard during the 'Dam Busters' raid on the German Ruhr Valley region in May 1943. The squadron leader radioed for the pilots to individually call in after the raid; those not calling in had clearly been shot down. But it was later discovered that one of the pilots who responded 'all okay' via his radio had in fact been killed as he approached the dam.

Phantom Second World War Lancasters have made occasional reappearances, as in October 1982, at the same place as one that had crashed in March 1945 in the nearby Bleaklow Mountains, killing a crew of six. A phantom USAF Dakota was seen in May 1995 that was about to crash-land. It later turned out to be the ghost of a Dakota that crashed on 24 July 1945, with the death of five passengers at exactly the same spot.

In another case, a Second World War Lancaster was seen from a car on the M62 near Saddleworth Moor, with smoke billowing from one of its engines, descending rapidly. When the car emerged from under a bridge the plane was nowhere in sight. In Sheffield Peaks in 1997 people saw an old prop plane flying low. They were certain it had crashed, but as usual no wreckage was found. The area was known to have had over fifty plane crashes during the war.[30] Many local people living near a collection of huts that was once Ridgewell Airfield during the war have heard the sound of piston engine aircraft, crashing planes and shouting airmen.[31]

Somewhat puzzlingly, Second World War phantom planes have been seen in Welsh skies rather than around airfields and hangars in England. A Second World War heavy transport plane was seen several times flying very low over Abergele. In 1987 it was described as being 'rusty' and 'very old'. A woman motorist driving near Wrexham in April 2009 saw a silent version of a 'huge' Second World War plane, flying low, possibly the same apparition that was seen in Abergele. But the Abergele version was large and unmarked, and hardly likely to be flown if it was a genuine antique, especially as it was seen doing dangerous low-level manoeuvres, barely 100 feet off the ground. Each time it appeared 'from nowhere', and then, of course, vanished as abruptly. In one case it dipped into a valley, but managed to pull out 'with great effort' before disappearing.[32] Another possible heavy transporter of this type was also seen in Brymbo, Wales, in 1977, again flying low. Local airfields knew nothing of these planes.[33] These impossibly low altitudes, which the RAF, air show organizers, or airport officials would clearly not allow, and the lack of markings, prove that these must be paranormal apparitions.

While filming the wartime film *Battle of Britain* in 1968 at the North Weald fighter base, which had been converted for filming uses, a phantom Second World War pilot was seen after the film lot had closed down for the day. He clearly didn't seem, from his demeanour, to be an actor, and in any event the actors were not in period costume at that particular time. He suddenly vanished when challenged. When the flyover of Spitfires was filmed for the movie, production staff said there were unaccountably seven planes rather than the six, and believed this to be last-minute change. But when the film was developed and viewed there were only six Spitfires.[34]

A man called Tony Ingle saw a silent wartime plane pass just *50 feet* above him in April 1995, 'experiencing difficulties'. He said: 'It was eerie ... I could see the propellers going round but there was no sound ... it was getting lower and lower ...' Some analysts have said this could have been the wartime B-20 bomber, which had four distinctive propellers. In Ingle's account, the plane seemed to disappear over a hedge into a field. But when he got there he could see only a typically quiet pastoral scene of a field dotted with sheep: 'Everything was silent, you could hear a pin drop.'[35]

A similar US version of wartime ghost planes can be taken with a pinch of salt. A worker for a building company in Long Neck, Delaware, in the early 1980s said he noticed an astonishing number of Second World War planes in the air that suddenly materialized from over the horizon. This was certainly a bizarre apparition, because no air show would display such a wide range of renovated planes from the Second World War era, and certainly not without any audience to witness the spectacle. The man noticed B-17s, B-24s, Thunderbolts, Corsairs, Hellcats and other army planes in the B category. Again they were flying at absurdly low altitudes, no more than 3,000 feet up. Although the planes would have been emitting a crescendo at that low height, he said he could only hear plane noise intermittently, and when they went behind clouds (apparently at 3,000 feet up!) the sound disappeared completely. He said the display went 'on and off' for about an hour. Unfortunately, no one else heard or saw this amazing air display.[36]

Other postwar apparitions of doomed civilian planes, accompanied by complicated sounds, visions, crashing noises, explosions and beacon lights, come from the USA. Sometimes people actually hear the crash, and see smoke and flames, and even smell spilt fuel. But upon investigation there is no sign of the wreckage, and no record of any missing or delayed flight is available. People in Westbrook, Connecticut, heard a single-engine plane on the morning of 15 January 1997 flying towards Long Island Sound before vanishing. The coastguard, rescue helicopters, fire departments and marine patrols searched the area but could find nothing.[37]

The corrobation of sights and sounds adds, of course, to the credibility of phantom plane reports. Three women in Ovando, 50 miles north-east of Missoula, Montana, saw an aeroplane trailing smoke and possible escape parachutes were seen, in April 1956. A boy in a nearby district thought he heard a 'big explosion' at the time, and saw red and yellow flames, but Malmstrom AFB officials at Great Falls and the Civil Aviation Offices agreed that no planes of any sort had gone missing.[38]

Another apparent corroboration of phantom aircraft sounds occurred in August 1976. A plane reportedly crashed in Reily Township, Butler County, Ohio, and in this case an amateur radio operator heard what he thought was a distress call, while a father and his two sons saw a white flash in the sky, then heard 'screaming' and a crash. But nothing was found, and airports in the area had no record of small aircraft filing a flight plan.[39]

Planes were also reported crashed in Dark Hollow, Pennsylvania, in November 1955. Reports of a plane flying at just

1,000 feet came in from ten civil defence workers. There was a possible explosion behind a hill, and flares were seen. Nothing was found of the plane, nor any wreckage. But the flares strangely persisted over several days. The aerial search involved nearly 300 firemen, police, civil air patrolmen, and civil defence workers.[40]

Some of these paranormal happenings hint at spiteful – indeed, unpleasant poltergeist-type – phenomena. In a few cases absurd and eccentric behaviour is reported, often out of keeping with the gravity of the scene being witnessed. In the Montana, 1956, incident mentioned above, a flat 'board' was bizarrely thrown out of the crashing plane. In another example, a year after the attack on Pearl Harbor, witnesses and radar signals could detect a plane approaching US soil from the Pacific coast. Military pilots were sent to intercept it. They saw an American P-40 that was clearly of the type stationed in the Pacific when the Pearl Harbor attack occurred, with obvious bullet holes and the landing gear sheared off. A bloodied and slumped pilot was seen in the cockpit absurdly 'waving faintly' and smiling at the other pilots, before 'crashing' to the ground. Once again no wreckage was found.[41]

In other ghost plane incidents, time warps are involved, with the same ghostly crash heard at different times, with signals coming from far away, and some people smelling the crash fires but not others. There can be tangled, complicated and dishonest accounts of the happening. Loud concrete slabs were heard crashing together, causing windows to vibrate, on 29 November 1996 at Miami Township, Ohio. The local police showed up, 'checking for an airplane crash'. But some doubt existed as to the time the crash was supposed to have occurred. Witnesses reported the disaster as between 7 p.m. and 8 p.m., but police said other reports were about an hour later. Other witnesses said they saw something with a red light. There was a massive search and rescue operation from several police agencies, spurred by an apparent beacon distress call received by a flight service station attached to Dayton International Airport. Strangely, this was not received locally but from an orbiting satellite! Further, the beacon signals varied over huge geographical areas, first appearing 15 miles from the Wright Brothers airport and then 2 miles distant. At about 9.17 p.m. rescue units began to smell hot burning rubber, but nothing was found. And there was a subsequent denial of certain reporting procedures by Dayton Airport officials that were later found to have occurred.[42]

Phantom ships

Phantom navies, as well as armies, have allegedly been seen. There was the sighting of an aerial navy by hundreds of people on a hill at Croagh Patrick, Mayo, Ireland, in the autumn of 1798. W. G. Wood-Martin, who wrote a book on the folklore traditions of Ireland in 1902, said this was the mirage of another fleet then in pursuit of a French squadron off the west coast of Ireland.[43]

Other nineteenth-century histories of Cornwall and Brittany in France contain references to ships seen to sail across land on anniversaries, or preceding storms or at the deaths of notable people. They were often seen at Land's End. Robert Hunt, author of *Popular Romances of the West of England*, 1865, described a ship of Porthcurno, coming out of the water and travelling a short way across the land towards Chapel-Curno, only to 'sail away' again. One William Bottrell, writing in 1873, repeated this, and said spectral ships were often seen 'sailing up and down the valley, over dry land the same as on the sea ...'[44] This legend was either attached to the ghost of a person who had lived more than a hundred years earlier, or was a 'token warning about sea-bound enemies.[45]

The 'Flying Dutchman' is the best-known ghost ship, named after the Dutch captain rather than the name of his vessel, which is seldom identified in witness reports. Captain Hendrick Vanderdecken encountered a severe storm as his ship rounded the Cape of Good Hope in 1680. He ignored the dangers of the storm and fanatically determined to press on regardless, but succeeded inevitably in sending all aboard to their deaths. As a punishment, runs the legend, Vanderdecken and his ship were doomed to plough the waters near the Cape for eternity. Sightings of the ghost ship occurred in 1835, 1881, 1935 and 1939. The last recorded sighting was in 1942.[46] The captain's log from the British warship, the HMS *Inconstant*, on 11 June 1881 was particularly revealing, as it described an incident that had also been witnessed by at least a dozen crew members: 'At 4:00 a.m.: 'The Flying Dutchman' crossed our bows. She emitted a strange phosphorescent light as of a phantom ship all aglow, in the midst of which light the masts, spars, and sails of a brig 200 yards distant stood out in strong relief as she came up on the port bow where also the officer of the watch from the bridge saw her …'[47]

The steamer SS *Valencia* sank off the coast of Vancouver, British Columbia, in 1906, after encountering bad weather. The *Valencia* eventually sank after hitting a reef, and only thirty-seven of the roughly 180 people on board survived. Sailors would claim to see the spectre drifting near the reef in Pachena Point.[48]

The P & O liner the SS *Barrabool* sailed from Australia to London on 26 January 1925. Officers on watch at night saw a strange light, after leaving Cape Town, through binoculars. They made out the hull of a ship, with two distinct masts, between which was a luminous haze. The ship was some 3 miles distant, and when approached she disappeared.[49]

Another phantom ship was seen on a trip to Galway, Ireland, from New York in the spring of 1932. Witnesses claim they saw a 'bridal barge of Aran', a ship dating back to the eleventh century, which was periodically, by all accounts, seen at night in the Atlantic. It was described as being of dazzling gold, and had red trimmings and bunting, all of which was highlighted by a full moon.[50]

The US Navy destroyer *Kennison* in the early winter of 1942 was patrolling off the Golden Gate on San Francisco's Pacific coast searching for Japanese submarines. Lookouts on the galley deck and on the gun deck instead saw a derelict two-masted sailing craft that had ploughed across the water within a few yards of the destroyer's stern. It was a shabby-looking vessel, and obviously dated from an earlier age – an apparition from the past. But it was seen only for twenty seconds or so before it vanished. There was no trace of it on the radar.[51] Elliott O'Donnell, writing in 1965, spoke of 'nasty sea ghosts' – in particular, of a ghost ship seen in Solway Firth, Cumberland, and the Goodwin Sands off the coast of Kent, where it is rumoured many ships have been sunk with all hands. Another writer, Gerald Findler, wrote of several people who have seen a ghost ship the *Betsy Jane*, sailing along the Solway, often about Christmas time.[52]

A 'ghost ship' was seen off French coastal waters in heavy seas and filmed on 1 April 2008 in long-shot, and is available on the internet. This ship was probably the St Pierre et Miquelon one, on which all thirty-two crew died in October 2001. The difference between the tranquil sea nearest the camera, and the bizarre heaving of the waves further out, was marked. The commentary did not make it clear whether the ghost ship was seen at the time the video was made.

Phantom tapes and radio messages

Over the years there have been experiments with 'ultrasound' frequencies that have indicated something paranormal is going on. Spectral voices on recording devices and radio transmissions are known to have occurred in many countries. Radio echoes from outer space have been heard on radios that could have been the voices of alien beings, according to some theories. In 1899 Thomas Edison, experimenting with a Tesla coil, said he could hear a series of clear dots and dashes on his radio monitoring equipment when it was turned off. These, he said, came from 'an unknown source'. Kenneth Corum, another electrical pioneer whom we met in Chapter 10, said these sounds could have been 'planetary signals'.[53] Guglielmo Marconi, in his yacht in the Mediterranean summer of 1921, also said he had detected 'unaccountable radio signals' which sounded like strange voices across the distant static.[54]

Unusual sounds also came from early radio receivers in 1927 and 1928, and again in 1934. There seemed to be some foreign language chattering, which was rapid-fire in an unknown guttural language.[55] An engineer at an Iowa radio station could hear 'musical beeps' bouncing between two points on a receiver. An 'alien voice' was also heard on a car radio in Jackson, Ohio, in 1964.[56]

A CD player began to make strange sounds while a mother and daughter were listening to it in a small Connecticut town on 19 April 2002. The music started skipping and electrical noises – including an animal apparently howling – played through the speakers. The other radios in the house gave out 'static'. Soon there was a total loss of power as a pulsating, colour-changing light appeared in the garden and brush area outside.[57]

During 'war games' at Warminster, Wiltshire, in 1966–7 tanks, light armoured vehicles and personnel carriers were put through their paces, keeping in contact with one another via army radios. But these seemed to suffer static interference and garbled signals from navy transmitters thought to be in the area. Brigadier Tony Arengo-Jones thought this might have been some kind of nautical trick! But the Admiralty records proved this was not so, although the actual navy signals received – according to naval listings of the relevant codes and frequencies – had been emitted from the Channel area five months earlier.[58]

Some apparently supernatural or abnormal events have been attributed to *infrasounds*. Infrasounds are known to produce feelings of unease in certain people. Vic Tandy, an engineer and computer expert, said ghosts could be 'very low frequency sound waves trapped inside buildings'. He attributed his own ghostly experience in a laboratory late at night to vibrations on a fencing blade he was working on that might have been gleaning energy from VLF sound waves or infrasounds. He said an extraction fan could have vibrated the air molecules at 19 cycles per second.[59]

The late biologist Dr Lyall Watson stated that the recording of phantom voices and sounds had been known about, and occurred, many times. He asserts there is no controversy about it. He reminded readers that in 1959 Friedrich Jurgenson, when recording bird-calls in Swedish forests, also picked up extraneous human voices. In the 1960s German psychologist Hans Bender at the University of Frieburg left a tape recorder on in a silent isolated room, and found it would contain faint imprints of recognizable human voices when the tape was played back.[60] Later Jurgenson, spurred by Bender's findings, linked up with a research physicist at the Max Planck Institute in Munich, and succeeded in picking up more voices on tape recorders left running in empty rooms.[61] Jurgenson was told by the Central Office for Telegraphic Technology in Berlin that the voices were real because they could also be reproduced on visual prints from the tapes. From the mid-1960s the Latvian psychologist Konstantin Raudive worked with Jurgenson, before setting up his own research project. Raudive would use a bank of up to four recorders which were synchronized together and protected by sophisticated instruments to block out freak pick-ups.[62] In one experiment in March 1971 over 200 voices were recorded, twenty-seven of them intelligible. By the end of the decade Raudive had built up an amazing archive of over 70,000 voices![63]

Lyall Watson says that anyone running a tape recorder with a blank tape attached to a simple diode circuit like a 'cat's whisker' in a silent, sound-proofed room can get similar results. He does, however, imply that entities or ghosts are aware of what is happening, and he says you need to announce in a friendly fashion that you are about to make a recording before you get results. On play-back, little can be heard at first, but 'almost everyone eventually hears voices' even though they have a 'peculiar cadence' which takes some getting used to. He says an engineer at a recording company installed special electronic filters to weed out the noises, but still heard them.[64]

Philip Imbrogno described a 'ghost box' that was apparently on sale in some cities in the USA for taping phantom voices. It is essentially an AM receiver which can scan up and down the commercial bands from about 550 to 1,640 kilocycles. Imbrogno says the more crowded the band with radio stations, the better it works. A clear standard signal is required to pick up 'spirit voices', because the strong signal lets other signals piggy-back off it. This often happens with weak radio signals riding on a stronger frequency. If you throw a 'dead carrier' (transmitting signals with no modulation), you can hear background stations hundreds of miles distant riding in with the carrier wave. Imbrogno says you can make a ghost box with an old AM receiver simply by cutting and reconnecting a number of wires in the back of the receiver.[65]

But critics inevitably point to the hazards of trying to detect phantom voices using radios and other devices that were capable of picking up fragmentary and distorted broadcasts, however they were tuned. A Cambridge scientist, Dr David J. Ellis, who listened to some of the Raudive tapes, said that listeners who thought they heard faint fragmentary utterances on a tape in a mixture of English and other European languages were probably catching the disembodied sounds of the disc jockey Kid Jensen who was broadcasting from Radio Luxembourg![66]

Other investigators, however, assert that it is possible that sound vibrations accompany EM anomalies, which might generate what scientists term 'Experience Inducing Fields' – EIFs for short. In the most haunted sites there was a continuous infrasound rumble, which may have been due to nearby traffic.[67]

Electrical engineers in 1982 working at a Welsh pub found they could trap supernatural sounds by inserting powerful electrodes into the inn's interior walls. By running a huge current of more than 20,000 volts through the building they could trigger electrons in the silica, and so release the 'trapped sound energy'.[68] Some brick walls contain combinations of silica and ferric salts, both of which are used in magnetic recording tapes. It was suggested that many extraneous noises were picked up by the recorders, including faint organ noises, a barking dog, and voices speaking in an old Welsh dialect. Jenny Randles, in her book *Time Travel* (1994), said that sandstone, millstone grit or rock with a heavy quartz content are particularly prone to somehow trapping haunting sounds.[69] Quartz crystals are known to vibrate and emit electrical signals when under stress. Renovation work, with hammering and digging work with pile drivers, can set crystals in motion and possibly reactivate a dormant energy field.

In an intriguing case, the voices of former residents of a house were heard in downstairs rooms in a small farmhouse in Somerset. Witnesses told of peculiar conversations that took place in November 1986, which all began and ended with a 'click' as if a radio had been turned on and off. A metallic click was also heard on a tape inside Point Lookout, a converted lighthouse at St Mary's County, Maryland, USA, in January 1973. This time, doors were heard to bang shut, along with sepulchral footsteps and the sound of furniture being moved. A tape machine recorded spoken phrases, some of which seemed to do with the treatment of injuries. Researches at local authority archives found that the home was used as a field hospital during the American Civil War.[70]

People in the Birmingham area in 1978 found a music tape wiped out and replaced with sounds of men screaming, crashing timber and rushing water. Local historians noted the house was above the site of a disused coal mine, and that a flooded mine disaster occurred there 100 years earlier. A researcher could hear, when investigating a humming case in Wales, with a recording device, low noises with a slight throb breaking the tone at intervals of about five seconds. At other times he could hear a bumping sound as if something soft and heavy was being dragged downstairs. Strange sounds and smells came from the attic.[71] Similar inexplicable sounds were heard during the night from a house in Sudbury, Suffolk, that were described as a 'high pitched continuous drone', and an 'electrical hum, like a whine', or an 'ethereal-like UFO'.

Another strange case involved the ghost of Richard Nixon. A security guard at the Nixon library in California, shortly after Nixon's death in 1994, noticed a phosphorescent green cloud floating above the former President's headstone. On another night he observed a figure entering the old Nixon clapboard house only to find the doors locked when he inspected it. On other occasions he would hear knocking noises coming from the museum's Watergate exhibit, and the next day the machines that played the White House tapes wouldn't work properly.[72]

In the meantime, the electromagnetic and microwave connection with ghosts was made by the Society for Psychical Research, in the USA, who said that they were in the process of being killed off by the mobile phone explosion that started in the late 1990s. The additional electrical activity could perhaps drown out subtle wavelengths that are detectable only by psychics who are sensitively attuned to the paranormal.[73]

Spectral wartime sounds

A former RAF policeman, Bruce Halfpenny, spent many years investigating haunted airfields. He said that at one old airfield, Kelstern, Lincolnshire, he could hear the sound of aircraft engines approaching the overgrown concrete runway at dusk, before abruptly cutting out. He could recognize one of them as a Lancaster bomber.[74]

Paranormal sightings and sounds have been heard for almost a century at Montrose Air Station, which is a heritage aerodrome open for tourists. There have been reports of eerie footsteps, ghostly figures and door handles turning. However, the strangest phenomenon is a seventy-year-old Pye valve radio that is on display in full working order. This has been said to pick up vintage broadcasts featuring Winston Churchill and the music of Glenn Miller, although the radio is not connected to any source of electricity. The broadcasts come on at random and can last for half an hour. The wireless broadcasts join a long list of mysteries at the heritage centre. As in other wartime military premises, there is also the sound of aircraft.[75]

In the early 1970s ghostly events were experienced at premises where Bircham Newton aerodrome in East Anglia was once situated. Originally constructed in 1914, the airfield had been abandoned in the interwar years, and then brought back into active service in 1939 as an air base for RAF and Commonwealth pilots. After the war, the base was converted into a hotel with squash courts at the rear, where a spectral Second World War pilot in uniform could be seen sitting about, watching, and then vanishing through walls. The squash players decided to leave a tape recorder in the court overnight, and later played back weird sounds, including groaning voices and the drone of an aircraft. This could have been the ghosts of three airmen accidentally killed during the war.[76]

The droning of aircraft can be heard at RAF Manby near Lough, in Lincolnshire. A Second World War pilot wearing the flying gear of the time and a long coat was often seen, and appeared whenever the droning aircraft could be heard. A recording made by a local radio presenter, Ivan Spenceley, in the museum hangar at RAF Cosford, Shropshire, picked up ghostly sounds of an ancient Lincoln bomber, the RF 398, that hadn't been flown for over fifty years. The sounds included the muffled voices of an air crew, droning engines that changed pitch, the operation of switches and levers, morse code blips and clanging hangar doors. Apparently spooky tales have surrounded this plane when it was first put in the museum.[77] Similarly, during the autumn of 1990 the uncanny sound of bumps, scratches, squeaks and girls' voices were heard, and there were sightings of a spectral air crew from the Second World War. This occurred within an aircraft hangar that housed a Second World War Lincoln bomber at RAF Cosford Aerodrome Museum. Staff at the hangar could hear the movement of switches and the rotation of wheels. One engineer saw a bright pinpoint of light at night that he could not explain.[78]

There is growing evidence that undersea sounds also have time-slip and paranormal causes, the ocean equivalent of phantom aircraft sounds heard at old airfields and hangars. It was suggested that mysterious rumblings from the ocean deep were echoes of Second World War naval battles. Weird sonar noises were picked up by US navy vessels operating in the Pacific during the year 1980. There were phantom explosions and gunfire. Scientists could hear garbled radio messages, both Japanese and American, and the names of dead sailors and lost ships could be discerned. Two years later similar spooky battle sounds occurred in the North Atlantic, but lasting over a longer period. These reports were regarded as credible, as the navy were using their own network of supersensitive hydrophones that were normally employed in the Sound Surveillance System (SOSUS), used mainly to detect the presence of submarines. Since 1952 the system had detected a plethora of different and inexplicable noises, and a top-level Pentagon report in 1982 insisted the sounds were of ghostly battles under the sea.[79]

Finally, there is evidence that some people can hear voices emanating both from the past and from the future, using future technology that has yet to be invented. The late British character actor, Wilfrid Hyde-White, a regular horserace punter, related the strange tale of a visit to the Royal Ascot racetrack in June 1935, a week before the races actually began. As he and a friend approached the rear of the stands they thought the meeting had begun a week early as they could hear the murmur of voices, 'hurdy-gurdy music' and the 'raucous tones of tipsters'. This was followed by the razzamatazz sounds of a royal procession making its traditional way down the course, as sovereigns have done since George IV's time. But judging from an invisible child's comment he assumed that the monarch in the coach was a woman rather than a man ('Look, the lady isn't wearing a crown!'), which would have been a spooky echo from the time of Queen Victoria rather than a future time-warp glimpse of the parade with George V in the carriage.

Then an even odder thing happened. Hyde-White heard a disembodied metallic voice announcing the runners and riders for the first race, as if from a tannoy. 'I can't remember when this system was first introduced in Britain but it was unheard of in 1935, and we listened in shocked amazement. More surprising still, the names of the horses were completely unfamiliar to us', he said. Then he heard 'They're off!', and the thunder of hoofs, then the final announcement of the first, second and third places. Not only that, but the tannoy announcer went right through the card, announcing the names of all the runners and riders. The pair guessed they were hearing the noise of a race meeting of the future, and made a careful note of all the winning horses. But, alas, by the time of his death in 1991 the ghostly runners had still not appeared before the Queen, and Wilfrid Hyde-White's greatest coup had not come off.[80]

14 Taking UFOs Seriously

Fearful speculation runs rife throughout the subject of UFOs. The entire gamut of 'LITS', 'UAPs' and 'BOLs' offer a wide range of interpretations. As we have seen with the orbs and ground-based lights, there is a great deal of confusing overlap between them and between marsh-gas type of lights, mirages and optical illusions, and 'spectral' lights. Unfortunately, UFOs might just be enlarged versions of these dreaded spectral lights, with all the ramifications of time-warps, ghostly apparitions and exotic energy physics that the following chapters will try to put into perspective.

UAPs, or 'atmospheric' phenomena, was an expression used by Jenny Randles, the writer on anomalist phenomena and UFOs, and the late Dr J. Allen Hynek, a well-known Nasa astronomer and UFO researcher of a generation ago. Astronomers also use the term AOP – anomalous observational phenomena. The term UAP was also used by the MoD, who also alluded to 'aerial phenomena' in their Condign report on UFOs, and emphasized 'atmospheric plasma' as one factor in the UFO issue. Various UFO investigation groups also use the term 'aerial phenomena'. A group in Derby used to call itself 'The Phenomenon Research Association'. But physicist Andrew Pike, who has often contributed to UFO journals, says that whatever acronyms you use in mainstream journals and newspapers, 'the stigma is too deep'.[1]

The fear that UFOs may be a manifestation of the paranormal originates from fighter pilots themselves. The strange apparitions known as 'foo fighters' were seen in the Vietnam and Korean wars by combatants on both sides, and by fighter pilots and bomber crews. In the Second World War they were described as ball-like, no more than 3 feet wide, and named 'foo fighters' because they were based on a comic strip popular at the time, called 'Smokey Stover'. In this cartoon the French word for fire (*feu*) was translated as 'where there's foo, there's fire'. But the foo fighters were not detected by radar and attempts to outmanoeuvre them were useless. They often flew in formation and stalked and accompanied the planes, sometimes back to base. Katie Hall and John Pickering say this was consistent with 'electrically propelled plasma concentrations', such a ball lightning.[2]

But some Second World War pilots saw a cluster of small silver discs in their flight paths, about 1 inch thick and 3 inches in diameter. These objects were gliding slowly, but when they collided with one B17's tail assembly, the discs 'went through them' in a paranormal manner without causing any damage. Data about the foo fighters was declassified from Air Material Command during the Second World War in the USA, as 384th Group on a final bombing run to destroy ball bearing factories in Germany in October 1943.

Optical illusions and varying misperceptions of eyewitnesses play a crucial role in many UFO sightings, a subject touched on in Chapter 8. One person may see a bright white light that appears to change to a bar of intense blue light almost 4 feet long. But observers on the other side of the light might see nothing at all. Aircraft lights can appear to be UFOs, flying away from or towards an observer, apparently hovering motionless. A civilian airliner from a certain angle can reflect the sun and seem to be a bright, cigar-shaped object without wings. Hang-gliders and microlights, swooping and diving with their triangular canopies, can reflect light in an odd way.

The space debris and meteorite explanations can swap places with one another within the first few hours of the object being seen. On the nights of 30/31 March 1993 two RAF police officers at RAF Cosford, Shropshire, saw two bright off-white lights above the airfield in the early morning darkness. The lights emitted a red glow as they disappeared over the horizon, just as if they were about to reduce their momentum as they penetrated the atmosphere. This event was confirmed by a Met officer at RAF Shawbury, situated nearly 20 miles from the Cosford base.[3] An early conclusion was that these lights were meteorite debris, but a later official one concluded that they were infalling space debris, possibly the re-entry of a Kosmos satellite.[4]

Further complicating matters from the 1970s onwards was the ever-lengthening variety of aerial weapons and flying craft that were undergoing testing and 'dummy runs', generating a growing number of fireball flashes, explosions and BOL apparitions, some of which I have referred to in Chapter 1. 'Flying triangles' have been seen in the USA since the 1970s, in particular over the Hudson Valley, in Pennsylvania and throughout the south-west of the USA.

Photos were taken on 23 January 1973 during a Skylab 3 mission of strange objects that Nasa claimed were satellites.[5] In August 1973 the USAF launched a Minuteman ICBM from Vandenberg aimed at the Kwajalein missile range, located some 2,100 nautical miles south-west of Hawaii. At an altitude of 400,000 feet, radar experts in the Pacific found they were also tracking a UFO next to the ICBM's cone, measuring about 10 foot by 40 foot.[6] Between 1975 and 1992 infra-red and optical sensors mounted on space platforms recorded 136 'fireballs' entering the atmosphere,[7] many of which were probably meteorites or comets.

Astronomers at the Kharkov University Laboratory in Russia on 15 May 1994 claimed they saw a 'white object' careening across the sky, and said it was a 'bolide sporting a tail of 180 miles in length', and that it flew north to south over the Kursk Belgorod and Kharkov regions at a speed of 20 miles a second.[8] It was only when the scientists discovered among other metallic substances in the debris a 'crumpled and exploded tube' some several feet across that they concluded it was probably a crashed satellite, and a pretty heavy one at that. Tests done at the crashed site said that the energy released was the equivalent of 1,675 pounds of trotyl high explosive.

In another doubtful case, a UFO was detected on 27 June 2006 close to the International Space Station. Space scientist Alexander Kireyev, in response to the suggestion that it was space debris, said it had no number in the space debris list. But it was indeed identified the next day as a USAF spy satellite called Hitch Hiker, launched as long ago as 1963, that whistled past the space station at 32,000 mph and passed within 1,000 feet.[9]

Many South American sightings are still attributed to space debris. In the city of Riolandia, Sao Paulo State, Brazil, on 19

January 2008, a huge UFO was seen during the night, and was said to have crash-landed and caused some crop damage. It was described as cylindrical in shape, and had oval endings. Dozens of investigators from the Association of Bioenergy Producers said no natural explanation could be offered for the way the crops had been bent although the object had brought about 'unusual crumpled vegetation [to occur] in some places of a vast sugar cane plantation'.[10]

Other sightings were swiftly clarified. The reported UFO seen in Russia on the night of 14 June 1980 was caused by the launch of Kosmos-1188. And the one seen on 15 May 1982 was a Meteor-2, and on 28 August of the same year it was a Molniya-1. On 5 November 1990 the re-entry of the Gorizont/Proton rocket body caused vivid flares across northern Europe, and the 28 January 1994 launch of Progress TM-21 from Kazakhstan was witnessed by airline crews who mistook it for a UFO.

A common explanation for very large UFOs is that they are dirigibles or experimental lighter-than-air craft. Many orange lights seen in multiple numbers are more often likely to be 'Chinese lanterns' – toy lights released into the air by party-goers – since such sightings of large groups of airborne lights are rare. The lanterns can cluster, and often seem to form triangles. They are the most pernicious factor when it comes to sorting out the genuine UFO sightings from the fake. However, these lanterns – most about 'wheelie-bin' size – contain flares that shed light no stronger than one 500 watt bulb, and they merely drift in the wind. They may slow down when the air currents falter, but they do not stop or speed up. For example, a photo of five bright UFOs was printed in the London *Metro* newspaper of 25 July 2007, outside a pub in Stratford-upon-Avon, with people actually pointing to them. These could have been Chinese lanterns, although we have to take the word of the witnesses who said the BOLs moved fast before stalling, then remained motionless for half an hour. Several of the lights were said to dart out of view at a fast clip.

The Stratford-upon-Avon lights should be compared with further reports of dozens of orange lights that moved slowly between 10.30 p.m. and 11 p.m. on 24 May 2009 in the Liverpool and Merseyside area. These were more than likely to be another example of Chinese lanterns, which were known to have been sent up after a late-night party. A photo of these lights, without accompanying explanation, was originally taken by Barrie Mills, head of the *Liverpool Echo*'s head of images, and appeared in the *Daily Telegraph* of 28 May 2009.

There has always been a mainstream, and commendable, approach to explain UFO sightings within the known Western scientific canon. Yet it is taken too far. Of Timothy Good, who has published extensively on the defence and security angle of UFOs, it was said by a critic that he 'accuses the authorities of a cover-up and yet either sidelines or ... ignores evidence that undermines [his] claims'. Philip Klass, in his book *UFOs: The Public Deceived*, said that a 1976 Tehran sighting of a UFO that disabled the avionics of two F4 interceptor aircraft, as well as ground control equipment, was merely the sighting of the planet Jupiter, coupled with pilot incompetence and equipment malfunctions![11] Jenny Randles, in the journal *Fortean Times* of January 2010, talked about 'Syndrome X', which means a tendency among many people to misinterpret phenomena. Yet in her books she writes that UFOs do have an external reality, and indeed has described them as 'missiles', as mentioned earlier in this book.

However, overt scepticism about the kind of explanation continually on offer is sometimes voiced by witnesses themselves and by the media. US air force officials were criticized for coming out with the 'Venus' or 'Mars' theory to explain virtually all sightings. These spokesmen lost considerable credibility with their version of the 'swamp gas' or 'fox fire' theory in the USA.[12] USAF spokesmen at Stewart AFB similarly 'blotted their copybook' and succeeded in discrediting official explanations for years by saying that the appearance of a red zigzagging UFO at Pompton Lakes, New Jersey, in early 1966, which was seen by police and others, was a 'special helicopter with bright lights'.[13] Newsmen said they were unwilling to accept that explanation, and did not report a similar sighting in October 1966.

The defence and space weapons angle

The postwar puzzlement over the new 'flying saucer' scares in the USA, with their mooted link with secret weapons, plus the lack of scientific expertise in judging UAP and space debris phenomena, has had a significant impact on defence and security organizations. The difficulty in correctly interpreting light phenomena spurred the North American Aerospace Defense Command (NORAD) system, with its specialized Space Detection and Tracking System (SPADATS), to use the expression 'uncorrelated observations'.[14] Running parallel to SPADATS is NORAD's Unknown Track Reporting System (UTR), who stated recently that 7,000 trackings of unknown objects had been recorded since 1971. But this mammoth monitoring regime, so similar to the space debris monitoring one, is plagued with duplication, overlap and double counting. Nasa, the US air force, various astronomers, and of course UFO monitoring groups – which are mostly populated with lay volunteers – are all involved with their different approaches and counting techniques. Nasa also collaborates on UFO tracking with the CIA, DoD, NRO, NSA and other intelligence agencies. Nearly all American UFO reports remain classified in spite of repeated attempts by 'ufologists' in the USA to get them declassified under the Freedom of Information Act, unlike the more open situation in the UK.

But there is a reason for the more secretive approach to US UFO research. Defence anxieties grew as Cold War tensions increased in the 1960s, exemplified by superpower confrontations, the growing stockpiles of US and Soviet nuclear weapons, and sundry spyplane crises. This made the further sightings of UFOs during ICBM testing in the 1960s ever more worrying. In the 1960s a shocked Col. J. J. Bryan III, USAF retired, said that 'hundreds' of military and airline pilots, airport personnel, astronomers, missile trackers and others had reported UFOs in the postwar years.[15] There were alarming reports of frequent air force, rocket, missile and naval bases being monitored or surveyed by so-called foreign spacecraft, as we shall see later.

Some notable US missile-shaped UFOs

1946   On 1 August, the pilot of a C-47 plane saw a cigar-shaped object hurtling towards him at 4,000 feet while he was descending to an airport north-east of Tampa, Florida. It swerved to avoid him, and he could see it was twice the size of a B-29 bomber and had 'luminous portholes'.[16]

1948 On 15 October, an American F-61 interceptor flown by a Japanese pilot in the newly created postwar Japan Air Force saw a 'rifle bullet' some 25 feet long, travelling at about 200 mph, which accelerated to 1,200 mph at an altitude of 5,000 feet.[17]

1948 In November, over Washington DC, two pilots had a dogfight with a glowing oval UFO at speeds of up to 600 mph.[18]

1952 On 12, July over Chicago, a captain and weather officer watched a red-tinted object with a small white body of accompanying lights make a 180-degree turn before disappearing over the horizon.[19]

1954 On 29 June, a crew of a BOAC airliner, with fourteen passengers, saw an object the size of an ocean liner accompanied by six smaller ones over Labrador. The objects travelled 80 miles alongside the airliner, a period lasting eighteen minutes.[20]

1957 A USAF RB-47 with a crew of six was followed by a UFO for over 700 miles for 1½ hours, as the plane flew from Mississippi through Louisiana and Texas into Oklahoma on 17 July. It was seen on radar and visually.[21]

1973 An army helicopter crew flying from Port Columbus airfield in Ohio were threatened by a red light that paced them and headed towards them at breakneck speed. They then attempted to descend to avoid a cigar-shaped metallic object with a dome, but they mysteriously found themselves ascending.[22]

1975 At Holloman Air Force Base, Alamagordo, New Mexico, in August, there were reports of a hovering metallic object 50 feet across, that had a high-pitched sound.[23]

1980 Fiery objects were seen in the sky by motorists in the suburbs of Houston, Texas, in December, and were chased by twin rotor Chinooks or were escorted by them. But the presence of helicopters suggested the object in this case was an experimental device that had malfunctioned.[24]

1981 An alert occurred when the US Navy's Atlantic Command Support Facility from Norfolk, Virginia tracked an unidentified object entering the Atlantic Area of Operations (AOR) in May. Warplanes were launched in pursuit from

Greenland and other northern Nato areas, and KH-11 satellites took still photos of a cylindrical shape that reflected sunlight, and that could outmanoeuvre all attempts to reach it.[25]

1990   A full-scale alert occurred when a US army reserve base on Juana Diaz on the southern coast of Puerto Rico was suddenly illuminated by a powerful white light.[26]

1994   On the internet can be seen a jet training programme video filmed at Nellis AFB, in southern Nevada, Las Vegas. A UFO with an appendage was clearly seen coming into view.[27]

Some startling allegations were made. The suspicion was voiced about unknown and menacing electrical power shortages that people in their cars experienced while witnessing UFOs, as many complained that their motor ignitions had stalled. The assertion that strange signals were interfering with the functioning of certain weapons systems also heightened concerns about advanced enemy aerial capabilities. A test explosion of a nuclear weapon of unknown origin mysteriously occurred over the Pacific in 1961 and caused a communications blackout over the area for hours. This was an apparent repeat of an event in 1952 when UFOs were seen over Washington, and were said at the time by the media to be 'extraterrestrials' who were responsible for a couple of US nuclear weapons that had been mysteriously destroyed.

The growing number of enquiries into UFOs inaugurated by the DoD and the CIA indicated how seriously the Americans have treated the subject, despite the often dismissive remarks made for public consumption. Indeed, several military and intelligence witnesses with the Strategic Air Command (SAC) and other nuclear specialists have testified to the reality of UFOs that had a non-naturalistic explanation. One was Ross Dedrickson, a retired USAF colonel who worked with the Atomic Energy Commission (AEC), in the 1950s. Later he was in charge of the Minuteman nuclear weapons programme, and had links with intelligence services. He gave evidence at a further UFO hearing in September 2000.

Electricity failures occurred in April 1962 at Stead Air Force Base, Nevada, after a large object had crashed near a power station in Eureka. Nellis Air Force Base admitted the object had been spotted on radar, and was chased by armed interceptor jets from the

Phoenix Air Defense command after being prompted by the National Military Command Center (NMCC). Various authorities, assuming the air force could not distinguish missiles from meteorites, initially asserted it was a bolide because it had a 'brilliant yellow flame'. Unfortunately for the theorists, the object took off again, proving that it was not a meteorite. It then headed over Reno towards Las Vegas, Nevada. But Dr Robert Kadesch, a physicist from the University of Utah, still insisted it was a meteor, saying that it exploded 65 miles up,[28] implying that the pilots were chasing glowing bolide fragments that had taken horizontal trajectory.

The meteorite explanation was again attempted in September 1964, when a UFO was said to have destroyed an Atlas-F ICBM during a test firing from Vandenberg. At a press conference in Washington in September 2010, when many military officers broke their long silence over the presence of UFOs during the Cold War era, former USAF captain Jerry Nelson described how a UFO pointed a bright light at the Vandenberg missile base.[29] Reports were classified as top secret and witnesses were told never to discuss what happened, despite the incident being recorded on 35-mm film. And at Niles, near Michigan, on 28 March 1966, a brightly lit 'spaceship', about 40 foot long, flew parallel to a highway.[30] This was observed by Federal Aviation control tower operators at Columbus and Atlanta, Georgia, as the craft continued on its journey across US skies. A lock-down scenario happened a few months later at the Minot Air Force Base in North Dakota when multiple observations of UFOs were made at three missile sites, and were again confirmed by radar.[31]

The most alarming event happened in March 1967 when an underground Launch Control Facility at Malmstrom Air Force Base in Montana spotted a UFO 'hovering' just outside the front gate. Former captain Robert Salas, who organized the Washington 2010 press conference, witnessed this incident. He said: 'The UFO was reported by my top-side guard. A bright red oval object appeared hovering outside the front gate.'[32] Other reports said that missiles in the Malmstrom silos began shutting down without being commanded to do so, and a combat crew commander said that it was extremely rare for more than one missile to go 'off-line' for any length of time, shut-down or not: 'In this case none of our missiles came back on-line. The problem was not lack of power; some signal had been sent to the missiles which caused them to go off alert.'[33] Speaking to journalists prior to the conference in Washington, Salas said: 'The US Air Force is lying about the national security implications of unidentified aerial objects at nuclear bases, and we can prove it.'[34]

During Congressional hearings on UFOs before the House Committee on Science & Astronautics in July 1968, physicist Dr James E. McDonald reported that the vast north-east power blackout of 9 November 1965 may have been caused by anomalous aerial activity, something hinted at in an earlier chapter. He said it was 'puzzling' why something had tripped the relay on the Ontario Hydro Commission Plant. Jimmy Carter, on 6 January 1969, in Leary, Georgia, along with twenty others, saw a UFO in darkened skies while at a club meeting house: 'It was big; it was very bright; it changed colours; and it was about the size of the moon. We watched it for ten minutes ...'[35]

During the two years from 1983 to 1985, during the 'Hudson Valley flap', a red light was often seen 'stalking' cars, and there were other ball-lightning type phenomena. Philip Imbrogno says there was an underground base in the Brewster area that might account for the River Valley sightings in Connecticut. The US government had bought land there, and many residents had seen military vehicles entering the dirt roads and never coming out. Helicopters were also noted to be landing in hills close to mine entrances.[36]

The mystery of the triangular UFOs

At a UFO conference in 1997 moderator Fife Symington, a former governor of Arizona, said he had witnessed a 'massive delta-shaped craft silently navigating' over a mountain range in Phoenix. It appeared to be solid, and 'dramatically large', with a distinctive leading edge with embedded lights. He said it was witnessed by 'hundreds, if not thousands, of people in Arizona ...'[37] From October 1980 to January 1982 in Arizona, enormous boomerang-shaped UFOs were seen by many witnesses. Sometimes they were V-shaped and surrounded by lights, and what appeared to be a leading 'searchlight'.[38]

UFO writers have gone into detail about these Arizona craft, which often appeared in dark military colours. When their routes were laid out across a map grid, the objects seemed to be moving from the north-western part of Arizona, over Phoenix and its surrounding area through valleys bordered north and south by mountain ranges and then back again along a well-defined corridor. One witness said they were huge and covered 'entire city blocks'.[39] Possibly black stealth transporters were designed not to return a radar signature and move at slow speeds, and were kept aloft, like blimps, by a buoyancy system. Two UFO journalists say it is possible that the US military did develop blimps that looked like black helicopters with silent rotors, or the air force could have been testing Lockheed neutral buoyancy aircraft, and other huge triangular-shaped transport aircraft.[40]

In fact, the 'stealth' aircraft explanation for UFOs made the defence and airspace issue more confusing and controversial, as it did even more so in Europe where the USAF jointly operated from Euro bases. In the Mojave Desert on 1 May 2003, a couple in a car saw instantaneously a large black delta wing at an altitude of about 1,000 feet. It was enormous – the size of a Boeing 747. But it could only be seen at a steep angle, about 45 degrees, but virtually in two dimensions because it was very thin. It glided uncannily smoothly, and there was no pitch or yaw. There was no engine on the 'wing', either; no nacelles, exhaust ports, no canopy or gondola, and no contrails. It was possibly a glider, but no flaps or rudders could be seen. Further, an obvious Black Hawk helicopter was keeping pace with the wing, and both were moving south at an incredibly slow 50 mph.[41]

More US triangular UFOs

1983   In March, New York State and Connecticut experienced a wave of 'boomerang' shapes at night, again of enormous size.[42]

1989   An oil rig worker saw a 'perfect black triangle' being escorted by two US fighters.[43]

2003   In Missouri on 27 May, at night near Joplin, several 'boomerang'-shaped white lights in a triangular formation appeared over a tree in a backyard. There was a dull 'reddish' glow in between, forming a circle with white lights making a meeting point.[44]

2003   On 1 August, passengers on board a pleasure cruise ship travelling from Tampa to Cozumel in the Caribbean Sea saw three lights in a triangular shape at night that hovered and flew away.[45]

2005   On 4 March, a flying triangle was seen at Westwego, Louisiana, that had white lights at each corner.[46]

2005   On 15 May, a massive illuminated triangle was seen at Carolina, Puerto Rico, with a shimmering outer glow.[47]

2007   On 23 November, a triangular object and a red ball were seen in the early morning before dawn, near Hollywood, California. The triangle had lights at the end, and changed its hue from red to green and yellow, and was 'very bright'. It was stationary for two hours before a red ball came into view.[48]

15 The Worldwide Enigma

As long ago as 1964 an official US agency called NICAP – the National Investigations Committee on Aerial Phenomena – published a 184-page document containing densely packed double columns. In it was listed 746 cases selected from over 5,000 signed reports which it claimed constituted evidence that UFOs exist.[1] Barely twenty years later the catalogue of cases listed in journals, government enquiries and books worldwide had grown phenomenally. In 1983 there was a 750-page catalogue of 'encounter' cases in France alone,[2] and a list of 200 landing cases in Spain and Portugal.[3] We can now guess that, in the second decade of the twenty-first century, many more UFO reports detailing not just thousands of UFO cases, but hundreds of thousands, have been published. In recent years there seems to be a rise in UFO sightings, according to media sources, especially from the UK and USA. The editor of *Nexus*, a magazine dealing with fringe science and suppressed discoveries, wrote in early 2009 that UFO sightings are on the rise everywhere.[4]

Further, the journals and reports – dating from the early postwar years – were not simply listed for the benefit of psychologists or cultural historians to mull over, and neither, in many cases, were the books. Many authors today, like Timothy Good, point up the military and defence implications of these UAP reports accumulated by defence, air force and intelligence agencies themselves.[5]

Indeed, initially, both Cold War superpowers and their allies logged UFOs as possibly suspicious enemy weapons, as we have seen in Chapter 14. Fears were aroused in the summer of 1961 when a huge 'disc-shaped object' appeared at an altitude of 60,000 feet, surrounded by a number of smaller objects, at Rybinsk, 93 miles from Moscow, where surface to air missiles were installed. Academics at the Soviet Academy of Sciences, and Colonel Boris Sokolov of the Ministry of Defence, admitted that many people saw a UFO over Petrozavodsk on 20 September 1977. Within a few months, the Moscow Air Defence network undertook top secret UFO investigations.

General Igor Maltseve, Russia's chief of general staff of the air defence forces, admitted that over 100 observations of UFOs by pilots had been reported to him during his term of office in the 1980s. In October 1982 UFOs were seen manoeuvring over a Russian missile base and had apparently manipulated launch codes.[6] Yuri Andropov, the Soviet leader from late 1982 to early 1984, ordered whole battalions and regiments to monitor sightings, and fleets of warplanes were launched at episodic intervals in pursuit of UFOs. A Mig-21 colour film taken in Soviet times, and put on the internet, showed a cylinder at first keeping level with the jet fighter, then shooting away at about 3,000 mph. It was a huge grey missile-like object about twice the size of the Mig. Late in July 1989 disc-shaped objects were reported by Soviet army personnel at a weapons depot and another military base in the district of Kapustin Yar, between Volgograd and Astrakhan.

But these objects were small-fry in comparison with some Soviet bloc UFOs. More so than elsewhere, they frequently assumed massive and threatening football-pitch dimensions. Most seemed to be metallic spacecraft or space stations of some sort. At a base in Byelokoroviche, Soviet Ukraine, a huge geometrically shaped object, an amazing 3,000 feet in diameter, was observed by large groups of military personnel on 4 October 1982, just sitting in the sky and occasionally rotating.[7] There was a similar disturbing incident in Krasnovodsk, Turkmenistan, in 1985 when an air defence radar station tracked an unknown object with a diameter of about 3,000 feet that was travelling at an altitude of 60,000 feet. A small disc flew out of it and, after several in-and-out journeys into the 'mother ship', the UFO finally disappeared into space.[8]

In September 1990 witnesses at a radar station near Kuibyshev, 490 miles south-east of Moscow, observed an unknown object which passed over an underground bunker. The radar signals seemed to indicate an isosceles triangle, about 375 feet long, with three whitish-blue beams of light coming from the front end. It remained stationary for ninety minutes and then took off into space.[9]

Another was seen right at the time of the collapse of the Soviet Union, on 28 August 1991, when a large cigar-shaped solid-looking object almost 1,800 feet long and 330 feet in diameter appeared over the Caspian Sea. It showed up on the radar screens of a tracking station on the Mangyshlak peninsula.[10] But when Soviet jets fired at it their electronics jammed up, as did their cockpit control panels and their radio communication links. But ground control tracked the object as it was reaching an amazing 42,000 mph while climbing vertically, and then dropping to an altitude of 14,000 feet.[11] The internet also showed a translucent blue-pink orb, hanging in space above Moscow on 1 July 2001. Another UFO was caught on a Russian security camera outside a residential complex on 22 May 2009. The object was travelling horizontally and had a luminous tail. In another internet clip a fleet of Russian helicopters was shown with a white orb passing by.

Captain Rodrigo Bravo Garrido, a pilot for the Chilean army, drew attention to UFOs in Chilean skies. Garrido was assigned by the army in 2000 to research aerial phenomena that could have had an impact on aerospace security. He said that the army had 'in its files many reports of UFO incidents from military pilots'. The Department of Civil Aeronautics also set up the Committee for the Study of Anomalous Aerial Phenomena, known as the CEFAA, to which pilots had to report. One incident in 1988, said to be 'fraught with danger', occurred when a Boeing 737 pilot on a final approach to the runway at the airport in Puerto Montt City, 'suddenly encountered a large white light surrounded by green and red'.

The Central Asian Republics also experienced massively sized UFOs, and the regional leaders in the defence ministries went along with the Cold War stance. In Turkmenistan in May 1990 military witnesses were surprised to see a giant disc-shaped craft, about 900 feet in diameter, reddish-orange in colour and with portholes around the rim, hovering some 3,000 feet over the town of Mary. Still in the Soviet Union at the time, air defence forces fired ground-to-air missiles at it, but the UFO apparently not only destroyed the missiles with 'intense beams of light' but also the pursuing jet aircraft.[12]

There were numerous sightings in southern Ukraine on 21 June 2003. An object was seen flying from south to north, from over the

Black Sea towards Moscow. It was a dull-grey cylinder 'with a bright front light and six dull-white windows'. The craft was about 600 feet long and 120 feet in diameter. A dedicated internet item on Cold War UFOs, 'UFO Encounters', says that nuclear missiles at a nearby Ukrainian base started to launch themselves in 2004.

Whether globular, disc-shaped or cylindrical, the Communist Chinese saw virtually all UFOs as 'secret weapons'. In October 1978 hundreds of Chinese air force pilots in the evening at an open air cinema show witnessed a massive oblong object, travelling near to the ground, that was shooting out white lights at its front end, with a luminous trail at the rear.[13] The pilots again spotted another UFO near the Chinese frontier in June 1982, and this was photo-illustrated in a new state-funded Chinese UFO journal of that time,[14] the existence of the magazine itself indicating that the Chinese took UFOs deadly seriously. A Chinese pilot in February 1995 was forced to take evasive action when unknown traffic approached him head-on while he was preparing to land at Guizhou provincial airport, south-west China. He was at a very low altitude of 7,200 feet and was starting his landing approach when his anti-collision system detected an onrushing object. He could see the object changing from a rhomboid to a circular shape, and changing colour from yellow to red.[15]

Radar stations in China's northern Hebei province in October 1998 also picked up a moving target directly above a military flight training base near Chanzhou City. It was seen by 140 people, many of whom reported that it had a 'mushroom shaped dome' on top with 'continuously rotating lights'. It was pursued by jet fighters to within 13,200 feet, when it abruptly shot upwards. Permission to shoot at the object was refused.[16] A UFO was shown on Chinese television screens on 11 January 2010, and was later shown on the internet. It was a brightly illuminated red tub-shaped object with fat sides, and had a green glowing rim. In addition, the Chinese videos on the internet showed a huge solid-looking flying pyramid UFO.

Other major officially documented sightings

1969 Turkey was inundated with reports of UFO sightings over the capital, Ankara, in October. One was a silvery oval, and the Turkish Air Force launched warplanes from Murtad air base.[17]

1974   A massive oval-shaped glowing metallic craft was seen on radar at the Air Defence artillery site at Binn, South Korea, in the autumn. It was some 300 feet in diameter, with red and green pulsating lights moving anticlockwise round the rim.[18]

1976   Parvez Jafari, a retired general with the Iranian Air Force, told how, on the night of 18 September, a UFO with multi-coloured flashing lights appeared over Tehran. Four other objects separated from it, an event that was witnessed by airport tower personnel. An alarmed Jafari and his crew launched themselves in fighter jets to investigate. Jafari tried to fire a heat-seeking missile, but his electronics 'blacked out'.[19]

1986   On 7 November, the Alaskan office of the Federal Aviation Administration reported that Japanese jets were chased by a UFO for some three minutes at flight levels between 3,000 and 35,000 feet. It was described as a 'huge ball with lights running around it'. The pilot said it was about four times bigger than a 747, and that it 'jumped miles in a few seconds', an event that was recorded on air traffic controllers' radar.[20]

1990   There were multiple sightings by military and civilian pilots of a massive 1,000 feet-long triangular structure with many lights attached some 15 miles east of Paris, possibly a continuation of similar sightings in Belgium. At one time it was flying parallel with RAF Tornado GR 1 jets over the North Sea. Explosions were heard in the Rheindahlen area on two separate occasions on the same night of the sightings.[21]

1995   An Aerolijheas Aergentinas flight, Boeing 727, with 102 passengers heading for the airport on 31 July at San Carlos de Bariloche, Rio Negro province, had to manoeuvre to avoid unknown traffic, which was described as an object with flashing orange lights with 'green lights at the end'.[22]

1996   On 10 August at Guarabira, Brazil, three flying triangles were seen at around 9 p.m. They were travelling slowly before zigzagging.[23]

1997   In Australia on 3 April at night at Umina Beach, near Gosford, NSW, a V-formation of UFOs illuminated 'triangles' of stealth-type aircraft was seen.[24]

2000 At Hammaguir, on a high plateau to the south of the Sahara Desert, a moving light, which had a 'cylindrical shape' and 'flames' of different colours, was witnessed by military personnel at a former French missile site in Algeria.[25]

2001 Dr Anthony Choy, a member of the Peruvian Air Force's official UFO study group, reported that on 13 October, in the northern city of Chulucanas, 'and before hundreds of people', eight spheres of red-orange lights appeared and were recorded on video. They hung in the air for over five hours, and moved 'in an intelligent way'.[26]

2002 The Peruvian air force announced the formation of its Office for the Investigation of Anomalous Aerial Phenomena (OIFAA) after several UFOs were seen.[27]

2004 In January, UFOs were seen over the Blue Mountains, west of Sydney, NSW, where there had been a wealth of similar sightings. Several people living in the Megalong and Kanimbla Valleys saw six orange-glowing triangular-shaped craft, perhaps 45 feet in width, flying from north to south.[28]

2005 On 24 June, UFOs were seen in the town of Yalapa, in the state of Veracruz, Mexico, by the governor, numerous police officers, and press reporters, and were photographed and filmed. 'Fleets' of UFOs over San Luis Potosi were also videoed, and still images were shown in *UFO Data* magazine.[29]

2005 A near-miss in July took place at an airport at Santa Roa, La Papa, Argentina. A luminous object was moving slowly and flying parallel to the aircraft, and was monitored by air traffic control staff. It emitted a brilliant flash of blue-white light.[30]

2007 A fireball was reported on 16 January by Fars News Agency of a UFO over Sepidar, Western Iran. It was observed for a whole hour, which rules out a bolide. It was yellowish, and had a 'reddish' tone in the centre.[31]

2008 On 1 June, over Rio Cuarto, in the province of Cordoba, Argentina, hundreds of local residents reported a UFO, which was also seen at airport control towers. The Argentinian air force put forward a spokesman to speak about it on a radio programme. Military sources co-operated with the Comision Nacional de Actividades Espaciales

(CONAE) to try to get a better understanding of these recurring aerial phenomena.[32]

2010 A pyramid-shaped UFO was seen over Colombia in March, and was shown on the internet.

2010 A pyramid-shaped UFO was seen in China, and shown on the internet. The video showed a smaller pyramid-shaped object emerging from elsewhere in the sky and heading for the bigger one. (Pyramid UFOs were also seen over the UK, France and Japan.)

European UFO reactions

The Europeans take UFOs seriously, and are prepared to be open-minded – even publicity-minded – about the phenomenon. There was a flurry of UFO sightings, peaking at the end of the 1980s, and alleged at the time to possibly be due to USAF experimental aircraft taking to the skies over the Continent. Wilfried De Brouwer was a major-general in the Belgian air force at the time that this famous wave of triangular or chevron-shaped UFO sightings took place. He participated in several television and press documentaries about the events. He said at a Washington conference that during the evening of 29 November 1989 in a small area in eastern Belgium, some 140 UFO sightings were reported. On three occasions the Belgian air force launched F-16 aircraft, the pilots of which noted the rapid change of speed and altitude 'which were well outside the performance envelope of existing aircraft'.

The Italians, unlike the British, have regarded UFOs as a military threat. A pilot of an Eagles Fokker 100 claimed his plane was buzzed by a large missile-shaped object on 25 June 2003 as he was due to land at Naples. The incident was immediately taken up by the National Flight Assistance Agency, the Ministry of Transportation Services, the Civil Aviation National Authority (ENAC), and others in Italy. The object was said to be white and reflective, about 20 feet long, flying at about 8,000 feet. The state prosecutor in Naples opened an investigation, but sinisterly confiscated the radar recordings of the event.[33]

The French regarded UFOs originally as a metaphysical problem, and were on the whole sceptical. The French Aerial

Phenomena Research Organization, with its French acronym GEPAN, was headed by many space scientists, including Claude Poher, director of the Sounding Rockets Division of the French space agency Le Centre National d'Etudes Spatiales (CNES). Between 1977 and 1985 GEPAN had logged some 1,600 UFO reports, with a surprising 38 per cent falling into the unidentified category,[34] way above the 5 per cent that British authorities usually admit to. In later years this French percentage was reduced to 25 per cent by Jacques Patenet, head of GEPAN.[35]

In March 2007 the French government released UFO files to the public, when the total of sightings had reached about 6,000, partly to assuage public concerns about secrecy and conspiracies, said Jacques Arnauld of GEPAN. In 1988, to take into account the connection with space debris, the organization was renamed SEPRA (Service d'Expertise des Phenomenes de Reentree Atmospheriques).

Dr Jean-Claude Ribes, a radio astronomer and associate director of the National Institute for Astronomy and Geophysics, pointed out that a secret report for the French government had been written on UFOs, mainly by military people from the French High Studies Institute for National Defence. This report, by all accounts, included analysis of both the defence and extraterrestrial hypotheses.

The British reaction to UFOs

Strictly, in the UK it was the Air Ministry that was responsible for collating and analysing pilots' UFO reports, and other aspects concerning flight path safety. Rudloe Manor's Flying Complaints Flight, concerned about near misses, was formerly part of the old S4 unit based in Whitehall.

In fact, the Airprox Board, the 'air miss' investigation branch of the Air Ministry, was kept busy throughout the Cold War years, as was the Civil Aviation Authority's Joint Agency Working Group (JAWG), although with little success at arriving at definite identifications of the flying objects.[36] National Air Traffic is the leading quango tasked with providing air traffic controls' guidance for fifteen of the UK's largest airports.

But it was the institutional link with the Ministry of Defence that was important. The entire UK intelligence and defence establishment up to senior Cabinet officer rank has spawned several overlapping official enquiry departments. A significant bridging link between flight safety and national air defence issues came about with the Project Condign of the 1970s, a formerly secret study that the MoD declassified and released in occasional instalments from 2006 to early 2010.

There was headline news in August 2010 when archives for the Second World War were released revealing a close encounter between an RAF aircraft and a UFO. People learned that a strange metallic object matched the aircraft's course and speed before accelerating away. Further, we learned that RAF jets were launched to intercept UFOs some 200 times a year during the Cold War.[37]

The 'flying saucer' thesis first gained prominence in the British media in August 1956. Two soldiers on guard duty in August of that year at Otterburn, Northumberland, saw a white glowing object over grassland covered with field guns, tanks and trucks. It ascended and descended as if it were surveying the landscape with intense lights for nearly two hours.[38]

It was revealed in 2010 that UFO phenomena filmed by RAF pilots as early as the 1950s was often secretly shown at MoD conferences.[39] MoD report revealed particular unease when UFOs with coloured light clusters were seen near RAF Bentley Priory, Hertfordshire, in 1984, with police teams blocking off roads.[40]

Kevin Patterson is an aviation reporter with a regular column in *Radio User* magazine. He received an anonymous radio recording from London Military Air Traffic Control revealing that pilots saw something at 3,500 feet. F-15 jets were then launched to investigate what the small black object was, and whether it was responsible for the 'flaming debris' falling from the sky over Dumfries and Galloway.

The concern about 'stray' cruise-type missiles in past decades was often successfully played down by Rudloe Manor officials. This occurred to me after discussions with a junior MoD official at the Royal United Services Institute in Whitehall, and at other times with an RAF officer from Rudloe Manor at a London officers' club. This spurred me to make more dismissive comments of ufologists than I should have done in my book *Sky Static: The*

Space Debris Crisis (2002). I alleged that too often people seemed to mistake UFOs for space debris or out-of-control military hardware. I wrote, 'the occult-like nature of the subject has encouraged scientists to give the subject a wide berth'. I praised the Russians for taking the matter seriously, even though they too concluded that most UFOs were space debris, and referred specifically to the breaking up of Kosmos, Meteor and Molniya satellites. I did point out that UFO writers such as Jenny Randles and Nicholas Redfern talked about 'cylindrical slender grey missiles' and 'dark wedged shaped objects' entering British air space, suspecting, like the authors, that these were unnatural phenomena. But I included them in my book solely to draw attention to the seriousness of the defence issues involved if the UFO angle had to be rejected. In one incident I mentioned that a 'dark cylinder' was observed by the crew of an Olympic Airlines Boeing 737 on 15 August 1985, while a cruise-type missile about 15 foot long nearly struck a Cessna flying near Sonherham in Sweden. A 'grey wingless projectile' was said to have zoomed past Dan Air flight 4700 from Gatwick to Hamburg on 17 June 1991.[41]

However, as there is now increasing public disclosure of UFO facts there seemed to be an attempt to revise the defence angle whenever RAF fighter jets were launched in pursuit of objects that *could* have been a security threat. Difficulties in perception arose in regard to 'pyramid' and 'triangular-shaped' UFOs. When a football-field-sized diamond object was filmed over Moscow in January 2010 and shown on the internet, it proved again that a rare phenomenon was seemingly becoming less rare, and that the Moscow apparition was likely not to have been faked as was earnestly suggested, because the images came from more than one source, and the object was separately photographed, and of course the apparition was replicated elsewhere.

Other major missile-type UK sightings

1956 RAF fighters from Lakenheath, over East Anglia in August, spent more than seven hours trying to shoot down up to fifteen luminous darting white objects which were picked up on army radar screens.[42]

1962 In May 1962, UFOs were seen by the crew of an Irish

International Airlines Viscount jet liner from Cork to Brussels, flying at 17,000 feet about 35 miles south-east of Bristol. The passengers also saw an object passing below them at 500 mph; it was brown and disc-shaped with eight antenna-like projections round the rim.[43]

1990   On 5 November, three RAF Tornadoes saw a UFO over the North Sea, while flying at 1,000 mph. Nato's air defence reported them on radar.[44]

1991   A passenger jet coming in to land at Heathrow nearly collided with a brown missile-shaped object passing overhead near Lydd in Kent.[45]

1993   On 31 March, there were numerous sighting of UFOs by the public and as many as nineteen police officers, across several western English counties, such as Devon and Cornwall, and over Wales. The MoD ruled out the presence of aircraft.[46]

2001   A metallic-looking projectile or snub-nosed rocket was filmed flying along the south coast, and a still was shown in *UFO Data* magazine of May–June 2006.

2006   On 1 July in the evening, two witnesses in Littleport, near Ely, Cambridgeshire, also saw a 'mysterious dark-brown crescent shape', about 1,500 feet in the air.[47]

2006   A 'near miss' took place when a pilot said he saw a 30-foot long black cylinder above his plane in Otmoor, Oxfordshire.[48]

2008   A helicopter carrying two police officers had to swerve to avoid a UFO with two continuous flashing green lights over Birmingham, in late November. Internet footage also shows a jet fighter pursing a UFO over Essex in the same year.

Major UK 'wedge' and triangular-shaped UFOs

1980   On 28 November at Todmorden, West Yorkshire, a police officer saw a diamond-shaped UFO with a rotating lower section. His car VHF radio and his handheld UHF radio both failed.[49]

1983   On 19 February, during the daytime, at Swansea, in South Wales, 200 witnesses saw a huge triangular-shaped UFO passing overhead.[50]

1989 An oil rig worker saw a 'perfect black triangle' being escorted by two US fighters.[51]

1990 A 'mysterious large diamond' hung motionless in the air for ten minutes next to an RAF Harrier jet over Pitlochry in Scotland on 4 August, and was recorded on colour film, before ascending at high speed, according to MoD and DI55 files, released in March 2009.[52]

1995 An MoD report cited in *UFO Data* magazine and several newspapers revealed that in January sixty Boeing 737 passengers from Milan, descending to 4,000 feet above the Peak District hills, saw a glowing wedge-shaped object from the aircraft windows. The pilot, Captain Roger Wills, said this illuminated UFO, covered oddly with a number of additional smaller lights, flew down his right-hand side in the opposite direction.[53]

1997 On 7 October, a gigantic flying triangle as large as three jumbo jets was seen over Hull. Thousands of lights were beneath it, all pale blue.[54]

2007 On 18 April, a ball of light was clearly visible over Dublin in the mid evening, and was moving silently in the direction of Dublin airport.[55] It soon turned into a huge chevron-shaped bank of misty orange lights.

The UK loses official interest

However, as time passed, 'UFOs', even if they were missile-shaped, gradually began to lose the interest of the MoD or the Air Ministry, as they had, already, with Rudloe Manor. The MoD in November 2009 declared that UFO reports were kept for only a month before being thrown out, and a month later it was announced that they had closed down their UFO files for good after more than fifty years of semi-serious investigations into the subject.[56]

But these files, dumped into various websites, and mostly from RAF pilot sources, were voluminous. One source said that between 1970 and 2000 there were over 700 UFO reports filed by UK police officers[57] (most police reports of UFOs are recorded in The Profus Police database). During the period of 1986 to 1992 there were continuing reports of phantom objects which were seen on radar and released in earlier MoD files on UFO sightings for this period. Within just a couple of years in the 2000s, some 7,000 reports were logged.[58] Some 394 mysterious hovering objects were seen in UK skies between January and the end of August 2009.

Even so, the MoD were convinced that, whatever UFOs were, they couldn't possibly represent a raised missile threat. There was a sense of resignation about the whole affair. UFO and UAP reports were distracting the MoD from defence issues. They said the 'UFO hotline' merely generated correspondence of no defence value. They repeated a well-worn mantra: UFO sightings had not revealed any 'potential threat' to the UK's air defences, ever, and that in hard-pressed times it would be an 'inappropriate use' of defence resources to pursue the matter further.

I have personal experience of this. The few pilots I have spoken to about UFOs say that although they may discuss sightings among themselves, they seldom, if ever, report them to the Air Ministry or the defence establishment, and here there is a marked difference between US and UK pilots. I recall seeing two white, solid-looking spheres flying parallel with a commercial aircraft as it flew low along the flight path of the Thames estuary at Erith, in Kent, towards either City Airport or Heathrow airport on 27 May 2005. I guessed these objects could have been no bigger than 8 feet in diameter. I contacted the Ministry of Defence about this, and within a week received a reply from a Mrs J. Monk at 'Freedom of Information-1', who merely confirmed that no pilot had reported any UFOs for that day, and repeated the standard MoD line: 'We are satisfied that there is no corroborating evidence to suggest that the United Kingdom's airspace was breached by unauthorised aircraft ... I should add that to date, the MOD knows of no evidence which substantiates the existence of these alleged phenomena.'

16 Are UFOs Aliens?

any experts, ufologists and even scientists do believe that UFOs are aliens themselves, or are a projection of aliens, or are robots made by aliens, or are vehicles that actually house aliens from outer space. On the other hand, the chronic irrationality of UFOs and their occupants, their dream-like and ghostly behaviour, clearly hints that they are not real 'beings' at all. They appear before farmers or stranded motorists, they dart around the sky allowing hundreds of people to see them, but show no interest in making contact. They are alleged to have performed primitive medical examinations on 'abductees' which no advanced technological civilization would need to do, at least not on that scale.[1]

Other ghostly analogies were made. One writer in the 1960s, Joseph R. Ledger, said that 'flying saucers' resembled wide-rimmed hats, and those that trailed appendages and appeared only at night heightened the ghostly similarities.[2] Others have noted the apparitional nature of UFOs, similar to ghosts that walk through walls, or the way people can be levitated at séances, and the way UFOs materialize from nowhere.

John A. Keel, in his *Operation Trojan Horse* (1971), said UFOs were possibly alien visitors from psychic dimensions. UFO sightings seem too often to be similar to the mind-created worlds that shamans encounter during their journeys through the subtler dimensions.[3] And the descriptions of 'aliens' who have emerged from UFOs, and who bizarrely approach and speak to witnesses, are too human-like, and they are clearly able to breathe our air, and show no fear of contracting earthly viruses, and display recognizable emotions on their humanoid faces, and speak our languages.

This is why many scientists, like Michael Talbot, defer to other-worldly or paranormal explanations for 'aliens'. From a lengthy study of archives and police reports undertaken by Lionel Fanthorpe and published in his *UK's Fringe Weird Report*, UFOs are the UK's most reported paranormal occurrences. UFOs have the ability to appear uninvited, as ghosts do in people's bedrooms. The *Fringe Report* catalogued more than 250 cases over the past twenty-five years, including a phantom hitch-hiker that was said to terrify drivers in Somerset, and a time-slip road in Lancashire where people travel back into the 1940s.[4] Jacques Vallée, a noted early commentator on UFO phenomena, and a physicist, says UFOs might be an old tradition in a new guise, resembling various folkloric traditions, from angels, elves, gnomes and supernatural beings from Amerindian legends.[5]

Many UFOs are surreal in character, singing absurd songs, or throwing objects such as potatoes at people. Tony Dodd, a well-known UFO investigator, affirms that BOLs are intelligent, but one of the criticisms of sceptics is the philosophical doggerel they come out with. Dodd tried to send a psychic telepathic message to a BOL flying some 30 feet above his head. All he got in reply was, 'I am the father of fathers, and you are the son of sons!'

Here I must admit that ufologists do not do themselves any favours. On an online version of *UFO Data* magazine, 'Andrea' says she dismisses other UFO websites 'who make the great information they have unreadable by including matter that could only be understood by rocket scientists'. UFO conferences often seem to attract people who belong to alternative lifestyle movements and New Age cults. Experts conducting seminars on UFOs are more likely to wear T-shirts, and sport tattoos and ponytail haircuts (while the spies in the back row continue to wear their suits and trenchcoats).

One UFO sympathizer, Scott L. Fenton, also pointed up the obvious problem with ufologists. They lack credibility, being 'weird anoraks that are easy prey for the corrupt authorities and sensationalist media ... [they] are our subject's greatest problem. It is their presence which feeds our enemies ...' Fenton criticized ufologists for expecting intellectually superior entities using force fields and laser beams to give us 'salvation from a dying world'. He also attacked the now defunct *UFO Magazine* for using advertisements for 'Grey Effigies', which other science magazines would not do.[6]

Astronomer Jeffrey Bennett, a specialist on SETI, has said it is 'almost inconceivable' that there are no other life forms in the universe, and he was prepared to mention the subject of UFOs in the title of his book *Beyond UFOs*, an update of four earlier editions of *Life in the Universe*.[7] In the 1960s Colonel J. J. Bryan III, of USAF, and a former aviation adviser to Nato, had stated his conclusions about the interplanetary nature of UFOs.

A major figure of importance in this controversy is astronaut Ed Mitchell: 'There's not much question at all that there's life throughout the universe. We are not alone at all ...'[8] Col. Gordon Cooper, a Mercury 7 astronaut, on a radio interview in Britain in July 2008, while not admitting he saw UFOs in flight, said that aliens exist.[9] The Japanese defence minister said 'there are no grounds for us to deny there are unidentified flying objects and some life-form that controls them'.[10]

The late astronomer Dr Carl Sagan said that Earth may have already been visited by aliens, and they may have bases in the solar system or on the other side of the moon.[11] Physicist Dr Paul Davies of Arizona State University also suggests that aliens could be right here on earth, unseen and virtually unknowable. He says it is 'entirely reasonable' to think that we share our planet with another life-form. This life might be lurking in poisonous lakes or even deep inside our bodies. Dr Lachezar Filipov, deputy director of the Space Research Institute of the Bulgarian Academy of Sciences, similarly said that 'aliens are currently all around us all the time'.[12]

A poll of 23,000 people in twenty-two countries found that more than 40 per cent of Indians and Chinese believe that aliens 'walk among us'. Europeans are more sceptical, with the number of believers being only one in eight, and most believers around the world are aged under thirty-five.[13]

Yet famed physicist Dr Stephen Hawking was quoted in the press in April 2010 as saying that humans should do everything possible to 'avoid meeting alien life forms'. Far from avoiding ETs, Professor Davies suggested that looking for weird forms of life living in extreme environments on Earth might be cheaper than going to Mars. Even some of the bacteria that scientists find difficult to grow and work with might be evidence of an alien biochemistry present on Earth, he suggests.[14] This point was made frequently by the late Sir Fred Hoyle in several books, especially in *Diseases From Space* (1979)[15]. Sir Fred suggested that organic life on Earth may have been 'seeded' by alien spores that drifted to Earth's surface via careening asteroids or comets. Other scientists point out that there is evidence of alien life in rocks that have mineral deposits inside them that seem to have been caused by different organisms with different metabolisms.

On the whole, however, the concept of alien intelligence is scientifically acceptable only if an out-of-Earth frame of reference is used. Sir Martin Rees, the Astronomer Royal, believes in ETs, but presumably not to the degree that would lead him to accept an alien explanation of UFOs.[16] 'Aliens', in other words, must remain in space and remain conjectural.

But even this kind of speculation is a threat to mainstream science. Belief in aliens undermines both neo-Darwinism and the 'anthropic principle'. The former says that pure chance and random shuffling of nucleic acids gave rise to intelligent life on Earth only, and did away with the idea of an intelligent creator. The anthropic principle – that Earth was uniquely placed in space to give rise to organic life – also undermines neo-Darwinism by suggesting some kind of pre-programmed cosmic design. Further, cosmologist Paul Davies says that consciousness in any being is not simply an accident arising from the random mutation that gave rise to brain matter, but is a fundamental emergent property of nature. He said consciousness arises from the laws of physics, quantum and classical, and from the way the universe self-organizes from the simple to the complex.[17]

Universal intelligence

Scientists often talk of aliens on other, older, planets having vastly more intelligence and technological capability than humans. The assumption is that greater neural complexity and species longevity alone make aliens more intelligent than humans. I do not think this is true, because intelligence is an abstract concept. Increasing neural complexity can take place, but intelligence cannot be 'increased' or 'decreased' in a quantitative fashion.

Many cosmologists believe that the universe itself is intelligent, and many parts of it individually consist of 'brains' of some sort. We return, ineluctably, to the 'hologram' concept of the universe. Whitley Strieber, who gained some fame for his books on abductee experiences, says that UFO encounters 'may be our first true quantum discovery in the large-scale world ... creating it as a concrete world'. Talbot cites scientists who say that UFOs are a 'holographic materialization from a conjugate dimension of the universe'.[18] This could be, he argues in his book *The Holographic Universe*, an accidental glimpse into the holographic record of the past.

The early twentieth-century astronomer Arthur Eddington said, 'We begin to suspect that the stuff of the world is Mind-Stuff'. By 'Mind-Stuff' Eddington meant that matter and energy are not only two sides of the same coin, but energy is also *information*. The logic underlying this is based on the idea that the structure of nature is founded upon the digitization of matter (particles) and radiation (waves). But when the electron considers itself a particle the chemical elements in brain tissue become *digitized* by the number of electrons in their orbits, or by the number of nucleons in the nucleus. Neurons (and connecting synapses) – which are the fundamental nerve cells in the animal brain – ultimately consist of electro-chemical networks that themselves are made up of electrons, neutrons and protons. This implies that the universe may be simply a gigantic electronic data processing system, and the human brain is hence a tiny working example of such a system with its brain neurons having been built up ultimately (as all living matter is) from cosmic rays, particles and energies.

As there are 100 billion neurons in the human brain, compared with about five billion in the brain of a small mammal like a dog, it would actually be unnecessary for any UFO brain to be more complex than the human brain. The Intelligent Universe, in a sense, created a brain of this complexity so that it could observe and speculate both about itself and the universe in which it finds itself. This idea formed the basis of the Anthropic Principle some twenty-five years ago, and pre-dated the controversial recent 'Intelligent Design' thesis, although the two theories are basically the same.

The computer is a prime example of intelligence being created from digitization. A computer deals in electrical signals, but electricity comes only in two sorts, plus and minus, or the types of spin that the orbiting electrons have around the atomic nuclei. In computers, patterns of electricity representing 'programmes' operate on the electrical 'patterns' in data memories, and thus transform the memories into *new* electrical patterns. That is all that 'intelligence' really amounts to. Indeed, Cambridge biologist Brian J. Ford suggested that even the neuron might itself be a tiny computer. He said that neurons cultivated in a laboratory emitted electrical signals of around 40 Hz. He added that they could address each other using 'discrete signals'. The brain hence becomes a 'vast community of microscopic computers'.[19]

On the domestic front, a new type of 'electronically enhanced wallpaper' can activate lamps and even control high-tech equipment in the home. There are such things as 'interactive walls', designed by Living Wall project, at the MIT Media Laboratory. Circuitry in attractive designs can be made with wafer-thin steel foil sandwiched between layers of paper which are coated with magnetic 'paint' that uses copper particles.[20] When combined with temperature controls, brightness and touch sensors, 'the wall becomes a control hub able to talk to nearby devices'; in effect, touching one device can activate another.

Insect intelligence

We now know that tiny creatures like amoeba and some insects simply don't need eyes at all, or even a 'brain'. Amoebas can take a different route to a piece of food or move away from some noxious chemical. Insect proteins known as cryptochromes can detect certain energy fields by making use of charged molecules called free radicals, according to Robert Gegear and Steven Reppert of the University of Massachusetts.[21] A blind male moth can recognize its mate by a single chemical signal, a pheromone. Some insects can migrate over long distances because they have in-built 'compasses' which they use to choose the best wind currents.

A honeybee has an amazing *one million* neurons, all operating in a collection of cells that would be no bigger than a grain or two of sand. Eric Warrant of the University of Lund in Sweden says that nocturnal bees and moths can see the night-time world in detail, and are aware of all the contrasts and colours in the real world that would be needed to find food and to escape from predators. Bees 'dance' to tell others about a good food source. They can also orient themselves via the sun's rays and the Earth's field to direct themselves to flowers, and their movements reveal the direction they have come from. Researchers report that bees can recognize faces, using a similar process that we humans use – by a system of configural processing, piecing together the components of a face, such as eyes, ears, nose and mouth – to form a recognizable pattern.[22]

There is extraordinary evidence of a psychic rapport that some bee colonies have with their human keepers that could not possibly be explained in biochemical terms.

Here are four amazing examples

- A woman described how her beekeeper father, in the 1930s, fell ill with cancer. A small swarm of his bees landed on the guttering above his bedroom window 'to keep vigil', and only went back to their hive after he had passed away.[23]
- Again in the 1930s, a beekeeper in Myddle, Shropshire, had thousands of bees turn up at his funeral held about a mile from his home, swarming inside the church and on nearby tombstones.[24]
- In May 1956, the death of a beekeeper in Adams, Massachusetts, saw thousands of bees converge around his cemetery even before the cortege arrived, landing in thick festoons from the pipe frame of the tent over his grave, even clinging to the floral displays. As soon as the burial was over the bees flew away.[25]
- When Mrs Margaret Bell, a beekeeper of Ludlow, Shropshire, died in June 1994, a swarm settled on the corner of Bell Lane for an hour, attached high up on a brick wall, during her funeral.[26]

The UFO brain

These remarkable empathetic events, often repeated, defy scientific explanation. The grain-sized brain of the bee seems to have psychic powers more often attributed to humans who have premonitions and out of body experiences possibly arising from the complexity of the human brain's neural circuitry. But this bee behaviour is eerie proof of the holistic nature of consciousness and the integrated universe where all parts, no matter how tiny, are part of a universal consciousness.

Similarly, evidence that UFOs could have primitive insect-like intelligence, and exist without 'eyes' (or the equivalent of camera-like lenses), can be found on the internet. Indeed, we can conclude that UFOs could quite easily perform their inquisitive aerial gymnastics with a 'brain' no bigger than that of an insect. Birds, like those that generally fly at night such as robins, thrushes and flycatchers, can actually see the Earth's magnetic field. The field might become superimposed on the landscape below – rather like the 'head-up' displays that pilots use in modern fighter jets. Aeroplanes use complicated electronics to operate the control surfaces and adjust the engines. They depend very much on computer brains that mimic and control behaviour, with radar and microwaves that can 'sense' things, and can set the positions of control surfaces to make sure the plane doesn't hit anything, a kind of primitive brain-like neural circuitry.

However, the notion that 'intelligence' could be divorced from organic brain matter was a common assumption among ufologists when they were in their more speculative and fluid phase in the 1960s and 1970s, after which a more formulaic stance took over the subject. The belief continues in most of the UFO movement that aliens are flesh and blood creatures, with vaguely humanoid appearances, and come from other planets or star systems. Dr Steve Dick, of the US Naval Observatory, says that ETs would have evolved into part-machine and part-silicon-organic robots, and have utilitarian appearances. They would not have to appeal to mates, and have attractive almond-eyed appearances. He refers to the 'intelligence principle' – in other words, when a species can improve its intelligence, it will do so.[27]

But there remains a residual belief in Earth-bound intelligent UFOs. Miceal Ledwith says UFOs might have some control over their own shape and form, while existing in a plasma state that utilizes energy from the wider cosmos. Many scientists and researchers attending a conference on orbs in Sedona, Arizona, in early 2007, also came to this conclusion.[28]

The most significant and detailed UFO film ever can be seen on the Nasa internet footage relating to the STS-74 and 75 Columbia shuttle. The crew were doing a 'tether' experiment, which was a 5-mile-long steel rope, which accidentally broke away from the craft and drifted in near space in February 1996 (according to one theory – the other says that the space shuttle crew temporarily lost the satellite and the tether, and went back for it). The six-minute video shows swarms of illuminated orbs, apparently consumed with curiosity, surrounding this tether, all showing animal-like activity. Many had notches, or a 'bite' on one side, with smaller bites on the other side seeming to shrink and fade. The orbs seemed to be the space equivalent of flagella, with their little 'heads' pumping backwards and forwards as they moved. The narrator at Houston made no comment, and another suggested the orbs were 'debris'.

That UFOs are illusive, and zoom away when approached by investigating military jet aircraft, suggests that they seem to possess some kind of primitive intelligence or self-awareness. In July 1952 two Maryland state troopers were driving through the town of Hebron, at midnight, and spotted a yellow dim light on the road, barely 5 feet off the ground. The object then kept pace with the police car, and just blinked out. Alerted, the officers arranged for a new crew for the following night, and the light appeared again at the same place. But as the men approached it on foot it moved away, before suddenly becoming invisible and reappearing somewhere else.[29] Later, it was revealed that this light had been seen irregularly in the same area for most of the twentieth century.

UFOs have even been reported as viciously retaliating against attempted attacks on them, as a provoked animal would. For example, a massive oval-shaped glowing metallic craft was seen on radar at Air Defence Artillery site at Binn, South Korea, in the autumn of 1974. Some 300 feet in diameter, it had red and green pulsating lights moving anticlockwise round the rim. It landed near the base, but the missile fired at it was destroyed by an 'intense beam of white light', as was the launcher. The object then disappeared rapidly from the radarscope, with a loud buzzing sound.[30]

Other UFOs seem to react as if they were vulnerable to human weapons. In March 1966 a US observer of a UFO at Bangor, in Maine, took a shot from his .22 magnum pistol at a UFO disc that had touched down near his car. He said the UFO had a clear bubble-type canopy on the top, and said there was a 'ping' as the bullet struck the object, making its lights flare up, after which it took off as if it were a wounded animal. The police subsequently found a scorched area at the site.[31] There was another case of ricocheting: during the 1960s at a farm in Rotherham, south Yorkshire, a pair of 'headlights' slowly made their way across a field. At the sound of a shot from a man carrying a sporting rifle, the lights sped across the field, through a hedge. Although the field was snow-covered, there was no later trace of tracks.[32]

But bear in mind the rather strange, poltergeist nature of UFOs. They often exhibit bizarre behaviour. In one dream-like event, UFOs appeared to witnesses as semi-solid, and transparent, with 'two men' inside, seemingly pumping a lever.[33] Dr Peter A. McCue, a clinical psychologist, wrote that cars and aircraft sometimes appear in apparitional form. He cited the case of a man in Glasgow in December 1983 who was waiting for a bus very early in the morning while it was still dark when he noticed an aerial object coming towards him at a low altitude. It was grey, shaped like a 'railway carriage', with crackling and humming coming from it. It had portholes and windows with yellow smoke swirling inside. But no one else seemed to notice it.[34]

In another strange case, in August 1992, the driver of a car travelling from Dunstable to Milton Keynes saw a bank of mist suddenly appear on the road, although other occupants of the car saw only a heavy downpour.[35] Paul Devereux, mentioned earlier in this book in regard to the 'earthlight' theories, referred to a huge black airship he saw, when he was a boy in broad daylight in 1954. It looked solid with the sun shining on it. It then vanished literally into thin air.[36] One writer says that as a child, around 1960, he and his parents saw a cigar-shaped UFO for close to three hours in the skies above his farm near Dalby, Queensland. He said there was no doubt the object was solid. Nine others watched the scene through binoculars.[37]

Can physics explain UFOs?

DAJ Seargent , in an interesting little UFO paperback published in the 1970s, made a curious observation that few others had made at the time, or even since. No one, he said, referring to the many air force sightings of UFOs, had gone to the trouble of producing a 'spectrogram' of a nocturnal light to tell whether it is from a solid or a gaseous source. This leads one to wonder why no pilot with a physics degree has yet declared over his Intercom: 'That UFO must be using at least one million kilovolts of energy to glow like that!'

Bright auroras and fireballs are a predominant feature of the universe and the solar system. Just a glance at *Astronomy Now* magazine or any astronomy website reveals a stunning translucent panorama of deeply coloured galaxies, nebulae and star clusters. A glowing ball of light should be seen against a cosmic background that contains trillions of BOLs.

Electromagnetism and the kilovolt electricity that gives us electric storms provides us with a clue. The bolt of lightning is actually visible as iridescence as the surrounding air is raised to a temperature of 20,000 degrees – about three times the sun's surface temperatures. But no pilot seems to have been able to compare the voltage of the UFO with that of the lightning-tossed storm he might have just emerged from. And what if the pilot's instruments told him there was no heat at all coming from the UFO?

There is evidence that some UFOs do emit heat. Over New York State on 1 July 1954, jets were launched to investigate a UFO, then the pilot complained of a massive stifling heat inside his cockpit. Two pilots had to eject. Four died on the ground when their plane crashed.[38] In fact, the UFOs' glowering aurora-type characteristics have profound implications for our understanding of energy and illumination. An unwitting clue as to how much energy UFOs use occurred on the evening of 17 March 1974 when electronic scanning instruments at the Manzano Lab section of Kirtland Air Force Base East registered a 'tremendous' burst of energy in the 250–275 MHz range. The burst was first noted in the upper atmosphere. A trajectory was plotted and a recovery team despatched to the small mountain community of Chilili, in New Mexico. The team discovered a metallic, circular object about 60 feet in diameter, which was moved into a hangar at Kirtland.[39]

The laws of physics are supposed to tell us that to *move* you must have energy, and to get energy you must convert mass. If you are using thousands or millions of kilowatts of illuminated power, that power must be gleaned from the Earth's atmosphere and environment, even if exotic physics is involved. If *no* known physics is involved, then this makes the idea that UFOs are mere holograms or apparitions more plausible.

We can perhaps only understand UFOs by relating them to quantum effects and other weird aspects of the atom and its innards. Is it possible that the multiple hues and erratic bursts of speed indicate that the UFO 'brain' is manipulating the intricate force-field of the proton, electron and the photon? The movement of electrons and protons in space, according to one view, functions in thin braided ropes or sheaths – the space equivalent of elastic tubes and spiral rings. They link up and create a massive web-like force that shimmers throughout the universe, and these sheaths bequeath an energy of nearly one giga-electron volts within the mass of the proton (GeV, energies of thousands of millions of volts). Science dictates that any proton with a kinetic energy greater than 1 GeV has more energy in its *motion* that it has bound up in its internal mass.

Intelligent BOLs could be using the voltage potential of the Earth's massive electrical field by liberating the energy from it. Physicist J. J. Hurtak says it is possible for advanced space vehicles to deploy an 'all-embracing energy field' involving an aggregation of earth-bound electrical and atmospheric and hydrological forces. Professor Hurtak refers to a 'perpetual pulse' that could aggregate all this cosmic energy.[40] Scientists Jack Cohen and Ian Stewart, in speculating about whether aliens exist on other planets, wrote about 'plasmoid' aliens 'woven from magnetic vortices; their "genetics" encoded in the topology of the weave. Magnetism is a major feature of the sun, and it "survives" the high temperatures and intense radiation without difficulty. Indeed, the solar environment is "just right" for magnetic aliens ...'[41]

Scientists insist that *dark matter* must exist to account for the massive gravitational pull observed over and above what can be accounted for by the known matter such as galaxies and stars that exist. Long-duration balloons have found an excess of high-energy electrons around Antarctica, which was said to be a telltale sign of the likely existence of dark matter. And if unknown energy, dark or not, exists, one can also assume that entities could be using this energy.[42] The late Professor Hynek described a UFO case of April 1957 that seemed to hint at some kind of powerful force field. Two women saw a UFO, a metallic machine-like tub about 5 feet tall, and heard nearby road signs rattling loudly.[43]

Incidentally, the accumulating evidence of UFOs in our skies confirms that the tub-like apparition is not at all uncommon, and competes with the usual 'flying saucer' and BOL image of the UFO. Some websites advertise a catalogue of over 340 photos of UFOs, out of an approximate number of 10,000 known to exist. Many are held by private researchers and organizations, and the *Daily Telegraph* also has online versions of UFO stills and video clips. The reader is particularly invited to visit the 'UFO Evidence' and 'Miscellaneous Nasa UFOs' sites. A blogger named Erik Sofge, in June 2009, showed UFO video footage of the STS-80 shuttle. Sofge pointed out the frequent comments made by the astronauts about UFOs seen out of the Shuttle windows, on a live feed from Houston, yet no comment at all came from Houston. In a long sequence of clips we can see that most of the Shuttle missions – a large number showing amazing footage ranging from missions STS-14 to STS-200 – reveal UFOs in the near distance. Other video footage, posted on the internet, showed objects seen by the crews of Gemini-4, Apollo 16 and Apollo 17.

The internet videos also confirm that many UFOs are indeed the knobbly and bulbous objects that people often describe, although most are circular BOLs or metallic-looking disc-shaped 'flying-saucers'. Some of the UFOs seen in space seem to have an organic shape, looking somewhat like sea horses, slugs or just strips of broken-off debris. On the STS-96 shuttle video we can see pictures of a bright blue piece of junk, looking like broken fenders or panelling, all startlingly illuminated. Some internet UFOs showed bright blue rods, like the Pipes of Pan, attached to an upright tower of sorts. Others were just clumpy groups of translucent materials, or artistic-looking 'lampshades' with curved bits attached to flatter bits, as shown on the STS-115-E video clip. Many were very clearly defined discs, with lights around the periphery edge, or underneath. Some were round balls, apparently solid, many moving. Some were shaped likes hats or smoke rings. Many of the Shuttle UFOs were luminous, and usually in the red-orange spectrum. Some were metallic, with sunlight reflecting on undersides, some zeppelin-shaped, some silver, some golden.

Unfortunately there are just too many of these UFOs, all with their different shapes and consistencies, actually to be 'alien space-craft', a point raised earlier. UFOs can adopt a bizarre superfluity of shapes, irradiated colours, appendages and beams of light. Some witnesses refer to UFOs having a top half and a bottom half revolving in opposite directions, hinting at 'nuts-and-bolts' type of spacecraft displaying mechanistic functions. Examples of objects taking on solid shapes was evidenced at Tyldesley, Manchester, in October 1968. A crude flying cross shape with wings coming out of its 'fuselage' at the sides was witnessed early one morning. There were two rows of porthole lights, and other lights flickered on and off.[44] On 22 July 2006 at Buckhurst Hill, a man saw from his garden a small dark-grey cigar-shaped object. But as it tilted it appeared to take on the shape of a silver doughnut. Then as it tilted again, the shape resumed that of the dark cigar.[45] On 18 April 2007, a ball of light was clearly visible over Dublin mid-evening. Then it turned into a huge chevron-shaped bank of misty orange lights. It was moving silently in the direction of Dublin airport.[46]

These appendage-type UFOs are remarkably clear on many videos, with the voices of witnesses expressing shock and puzzlement. A long projectile with odd struts projecting below was seen over Toulon, France, on 8 August 2009, sailing close over rooftops. A still photo of a bulbous, black object with 'attached irregular things' was seen over Grenada, in Spain, on 19 May 2006 and was shown on the internet. A Polish UFO, seen at Miedzyrzec Podlaski on 8 January 2006, had a curious bulbous, metallic shape, with a rim, and appeared in a high-resolution picture in *UFO Data* magazine. It appeared to be hanging barely 100 feet above a field. The editors of the journal seemed to think it was genuine, because the sunlight reflections seemed to be right.[47] (Other UFO pictures have been rejected as emulsion blemishes or technical faults precisely because apparently solid objects do not show light reflection or shading.)

An elongated UFO was videoed at a busy campsite in Brean, Somerset, in the summer of 2008. In it the cylinder, possibly several feet long, apparently solid and perhaps metallic, can be clearly seen against a blue sky, with the sun glinting off its top edge. This video appeared in public in February 2009 after it was sent to a website listed as www.burnham-on-sea.com. It was reproduced as a still photo in the *Daily Express* of 12 February 2009. The cylinder can be seen darting up and down for ten minutes. BBC newsreel footage of a military training exercise also inadvertently showed a jet fighter in the skies with an elongated dark object overhead, and this can be viewed online at the BBC News website or via YouTube.

The long dark cylindrical UFOs – reported by witnesses who are unaware of the sightings of others – also often seem to have the same tiny nozzle at the end, and this is a feature that can be clearly seen on many internet images of 'flying cylinders'. A long cylindrical glowing object was seen and videoed over Salt Lake City on 13 June 2007. Again the pinch could be seen at the end. The notch syndrome is also seen on video clips, and in still digital photos of 'orbs' at family gatherings, discussed earlier in this book. These repeated idiosyncratic characteristics could not have been jointly or conspiratorially agreed on beforehand by a varied class of witnesses, space scientists or photographers.

Time-warp UFOs

The existence of rotating discs also hints at the manipulation of velocity and gravity, and the torsion energy of particles. One can imagine a super-density disc with the strength of its gravitational attraction near the centre. The rotation of a disc causes an outward centrifugal acceleration, and this causes a horizontal acceleration that everywhere increases linearly with distance from the axis. If this disc were somehow supported in one Earth gravity field of the real Earth, then on top of the disc there would be a gravity field of two Earth gravities. On the bottom of the disc, near the centre, the one Earth gravity attraction of the mass of the disc would cancel the one earth gravity attraction of the mass of the Earth. 'Earth tides' could be involved that somehow succeed in nullifying gravitational forces, with the rotation of the UFO (one of the most common descriptions), causing outward centrifugal acceleration. The late physicist Robert Forward reckoned that once anti-gravity has been mastered using earth tides, the horizontal acceleration of the object could occur at vast speeds. He said we could, for the time being, ignore Newton's theory of gravity and use Newton's theories of motion instead.[48]

Other studies explain not only how energy could be tapped from the universe, but also explain the time distortions that UFO witnesses experience when they accidentally get close to the time-travelling UFOs. Scientists who believe in the 'multiverse' say that there are an infinite number of bubble-like universes attached to ours: 'At any time two or more of these parallel realities may merge together and allow matter to stream from one to the next.'[49] For example, two men in a car heading down a deserted country lane near Tarbrax, West Lothian, on the night of 17 August 1992 came across a black 'two-tiered disc-shaped object' hovering just 20 feet above the road. Frightened, they decided to accelerate underneath the object rather than back away. But they were then surrounded by a 'wavy black mist' that came from the object. When the men arrived at their destination, a short trip that would have taken fifteen minutes, they found that one and a half hours had mysteriously elapsed.[50]

In another case, a twenty-one-year-old man who was driving at night through a Northamptonshire village in September 1973 suddenly saw a bright light heading towards his car. Within seconds he found himself wandering along rural roads near Bedford on foot. It was daylight and he was soaking wet. His car was later found some miles away in a village field surrounded by mud but with no tracks leading to it. In January 1992 a young woman was driving on a road in Szekszárd, Hungary, when a red glow (some writers say a 'white light') swooped towards her, after which she miraculously found herself in a field totally covered by snow, but again with no visible tracks and despite there having been no snowfall for some days.[51] A glowing object was once seen from a car travelling towards High Bentham in North Yorkshire. But again there was a time lapse, as the occupants of the car suddenly found themselves approaching another town off their intended route. Retracing their journey later they found that the journey only took nine minutes, but the journey on the night seemed to take much longer.[52]

Weird time-warp UFO cases

1959 A businessman driving to Bahia Blanca, in Argentina, was surrounded by a white cloud that descended out of nowhere. When he regained consciousness he found himself on a country lane some 800 miles distant but without his car.[53]

1968 In Argentina, a couple in a car vanished while on the road to Maipu, when a dense fog surrounded them. After a time lapse of forty-eight hours and a period of unconsciousness, they found themselves some 400 miles distant.[54]

1973 A motorist encountered a glowing mass when approaching the village of Little Houghton, Essex, at night. He lost consciousness and awoke to find himself some 20 miles away in the early morning, soaking wet even though the weather was dry. His car was located in the middle of a field, locked, without no trace of tyre tracks.[55]

1974 In October, a family in a car found themselves enveloped in a 'green fog'. There was crackling on the car radio, and they had the sensation that they had bounced over a hump in the road. Two hours had unaccountably elapsed.[56]

1980 Near Todmorden, in West Yorkshire, a police patrolman spotted a rotating mass on the road causing bushes to vibrate in the wind. Again another time lapse occurred and the policeman awoke to find himself in a deserted road.[57]

1992 A family driving to Milton Keynes in August encountered a thick mist that surrounded their car. After a momentary time lapse they found themselves some 8 miles from where they had been.[58]

UFO colours and radiances

The most obvious features of UFOs, let us be clear, are their brightly illuminated changes of colour. So let us refer for a moment to the elementary particle of which everything in the universe is made, even light itself. The importance of the *photon* to our everyday world can be demonstrated by our television and computer screens. On your computer you can click your mouse on to an icon parked to the left of your screen. You can then drag a dimmer, translucent version of this icon across the screen. When the mouse click is released the icon is on the other side of the screen, leaving a spectral gap in place of where it was earlier. The icon itself, consisting of photons of light, travels across a sea of pixels that consist of other tiny photons of light. So the glowing ball of light – like the pixels on a computer screen – would be seen against the background, which is also a potential ball of light. The air consists of subatomic particles that are virtually identical to the photon in that they are part of the EM spectrum. As an accumulated mass of trillions of photons, the UFO could 'transmit' itself across the sky in an instant. It could be simply displaying itself as a 3-D image, consisting of nothing more than a large ball of plasma traversing this complex grainy particle soup.

Now, an important clue to UFO changes of colour comes from the knowledge that a radiating object glows redder when it is put in a stronger gravitational field and bluer when it is placed in a weaker field (although this is not the same explanation that gives rise to changing colours via atmospheric drag). Many circular, disc-like UFOs, because of the way primary colours seem to be rotating and oscillating, seem to be going through the equivalent of *chemical* changes, especially as they rotate from red to blue and back again. In experiments, chemicals can change colour by adding and removing oxygen to create a cyclic oxidation/reduction reaction. Chemicals, with further tinkering, could conceivably expand outwards as rings alternating between blue and red. Clearly UFOs are not undergoing chemical reactions, but the photons they are made up of, and additional inputs from electrical or cosmic field forces, could possibly make UFOs go through a similar routine.

However, as scientists are also able to convert signals back and forth from the optical to the electrical, then similarly UFOs could use gravity fields to make magnetism and electricity switch places with each other. Experiments in LED technology have revealed how subatomic energy can be gleaned at specific wavelengths to produce brilliant light, especially *green* light. Todd Pedersen from the Air Force Research Lab in Massachusetts, and Elizabeth Gerken of Cornell University in Ithaca, New York State, have created emissions of 'unprecedented brightness' as well as 'bright green speckles' by using pulses of radio waves sent from a base stationed in Alaska.[59] An LED-emitter contains a phosphorescent chemical that absorbs a portion of one colour and re-emits it as another.[60] On 28 May 2003, a man outside his home in Omak, 125 miles north-west of Spokane, Washington State, spotted a 'glowing green ball' heading north towards the Mackenzie Mountains and Alaska; this was followed forty minutes later by another green orb, but from a different trajectory.

The combination of a *green* UFO with an apparently solid exterior, but shooting out blue flames, is definitely puzzling. Robert Stanley, a Washington photographer, recorded *green* glowing objects over Capitol Hill in July 2002, with some leaving a *blue* signature. Sometimes the objects turn green when approaching aircraft, or leave a tinted vapour trail. Extraordinarily blue UFOs were shown vividly in video footage over Selby, Yorkshire, in June 2006, and were reproduced in *UFO Data* magazine.[61]

In the meantime the proliferation of beams, multiple lights, the glowing and rotating parts of the UFO, might be a manifestation of what happens at the photon and the proton spin levels, with a lot of overflowing light energy leaking out into our skies. UFOs, almost accidentally, might be using too much light energy for their purposes, due to some bizarre manipulation of cosmic forces.

Science can perhaps provide a vague clue here. Niels Bohr, another famed father of quantum physics, guessed that the spectral line in atoms must correspond in some way to the energy of the electron in its orbit. As an atom is heated its electrons jump from one orbit to another, higher and higher, and at each jump it gives off energy which appears as a line on the spectrum. The electron can then be made to drop a voltage point, or jump to a higher energy level, leaving behind positively charged micro-spaces known as 'excitons'. The energized electron gives up its extra energy and plonks itself back into the exciton. The extra energy is revealed as another photon. The photon is thus some kind of a 'force carrier' for the subatomic world, and can be seen everywhere at the microscopic and macroscopic level. A chain reaction can occur in a billionth of a second, while energy begins to cycle between light and matter. It soon becomes impossible to tell in which of two states – matter or light – the photons, electrons or excitons are. It is precisely this enigma – whether UFOs are BOLs or something more solid – that ufologists are confronted with.

The stray shooting beams of light, often reported, are invariably accompanied by other seemingly redundant light energies apparently assessed at tens of thousands of kilovolts. At Mexborough, South Yorkshire, late in the evening of 7 September 2007, a stationary light in the sky was 'giving the appearance of shooting sprays of light'. This object was photographed digitally and reproduced in *UFO Data* magazine, looking like a dimly glowing disc of light speckled with a tapestry of coloured fragments,[62] with additional greenish-orange 'piercing' lights, plus a number of small lights on its outer rim. There is an eerie similarity with the orbs of much smaller dimensions described earlier in this book.

In May 1977 a crew of a Vulcan bomber reported a spectacular encounter with a UFO at 43,000 feet, over the Bay of Biscay, that resembled an array of landing lights. But the object also shone a 'long pencil beam of light ahead'. The light then appeared to go out, leaving a diffuse orange glow with a bright fluorescent green spot, and the UFO was sufficiently solid and reflective to be recorded on the bomber's radar. Often UFOs are seen on radar only, imitating the outlines of aircraft, but invisible outside the airfield. In the mid 1960s a US military unit noted a slow, nocturnal UFO that glowed red. It was observed on two radar screens; one was the missile-control, and the other the search radar. When the UFO changed its hue to blue it was seen only on one screen. When it reverted back to red it was seen on both radars.[63] On 22 May 2004 in the afternoon, at Nelson, Lancashire, a black 'stick' was seen in the sky, turning slightly. This apparition lasted about thirty minutes, abruptly changed direction, then became 'shorter' before it vanished.[64] On 21 August 2005 at Wakefield, in the evening, a witness saw a black object constantly changing shape, and also vanishing.[65] In Rainhill, Lancashire, in January 1972, a bright light shrank and enlarged, switching on and off, and changed colours repeatedly from the usual red, green and blue.[66] The same object was seen on 17 November 2007 at North York, Toronto, Ontario, at about midnight, and was also shown in a *UFO Data* magazine picture.

UFOs displaying colour and shape changes

1957 A colour-changing UFO was seen by a Portuguese jet fighter at 25,000 feet between Granada, Spain and Portalegre, Portugal.[67]

1978 A man reported that a tank-like object with coloured lights on its top, on playing fields near his home in Huntingdon, Cambridgeshire, descended on 5 November. It had a 'sort of telescopic probe' protruding from its metallic, dome-like structure which appeared to search the ground. It had weird attachments that swivelled, and had small wings like stabilizers that were constantly in motion.[68]

2003   An object first appeared as a white light in an unidentified region of Turkey, which then turned into multi-colours. In another incident, on 10 August, some fourteen UFOs were seen at night, and again were multi-coloured. They flew in formation, with the white lights at the front and the red lights at the back. But several of those at the back were cigar-shaped, while those at the front were globular.[69]

2005   In June, in St Petersburg, in Russia, a UFO formation in the form of a Christian cross was seen. It had perfect 'white orbs' popping on one at a time, then the orbs went out suddenly with a silent 'flash'.[70]

2008   On 11 June, in North Yorkshire, a witness saw a 'cork-shaped object that glowed like an angel', and that flew over some trees.[71]

2008   On 8 September, an object was seen in Blackpool. There were circles of light beneath it that emitted a dull orange glow. It was about 150 feet long.[72]

Wavelength-trapping techniques mean that the radiation frequency of the particle can be manipulated, which means in turn that on the larger scale the colour spectrum can be subtly altered. Light can be curved inward and absorbed on the local level as it can in space,[73] according to scientists at Purdue University in West Lafayaett, Indiana. This could be done with amazingly simple electrical devices tuned to operate at subatomic levels. Indeed, light can actually be stopped, at least experimentally, when semiconductors made out of 'quantum dots' can cram electrons so tightly into a microscopic device that they have no dimensions to move in at all. These quantum dots act like mini black holes where the gravity at the atom's centre – the powerful forces around the nucleus – can prevent light escaping.

LED technology can also be manipulated to do the same. In 2007 Ortwin Hess of the University of Surrey showed how he could trap light in an LED-based device that could 'taper' a waveguide – which is a structure that guides light waves at the terahertz frequency range down its length. As the waveguide became narrower, each frequency of light came to a stop at a different place. Other experiments show similar effects. Different frequencies at which atoms reveal themselves can be trapped at different spots along a 'grating' device of certain lab experiments.[74] Hence if UFOs are gleaning motive power from Earth's gravitational forces, there might be a reason for them to allow 'frequency trapping' to occur, and hence for its colour-scheme to change.

The UFO spectrum contrasts are of course mystifying. Multi-hybrid UFOs – partly atmospheric phenomena, partly paranormal and partly 'flying saucer' – indicate not only the ability to 'morph' from solid to less solid. Even soft objects, of course, remain solid, although the fact that many UFOs are insubstantial helps to explain their silence. The pilot of the Boeing 737 in January 1995 said to Airprox, the 'air miss' unit of the Air Ministry, that there was no sound or displacement from the UFO encounter.[75] Some UFOs even seem to be able to pass through or dissolve into other UFOs. They often appear as coloured BOLs and simultaneously as 'flying saucers'. UFOs perhaps could go through a solid phase by flying through an unknown region of space where particles can alter their states from massive to massless. Sometimes witnesses see these changes in a UFO's solidity, with the craft or object becoming less solid before their eyes. Indeed, this is a very common observation made by witnesses. Of course, we know that subjecting objects to heat and cold effects can make them 'morph' from solid to less-solid. Some UFOs seem not just to become 'less solid', but seem to turn into a kind of translucent sponginess.

A commander in the Peruvian Air Force, Oscar Alfonso Santa Maria Huertas, reported an incident in April 1980 when he tried to shoot down a large balloon-like object, which was reflecting the sun at 1,800 feet, in a restricted airspace. The projectiles had no effect upon it, as if they were *absorbed* rather than bounced off it. On approaching the object, Huertas saw it was about 300 feet in diameter with a 'enamelled, cream-coloured dome'.[76]

How could UFOs become invisible?

Let us, in the dying pages of this book, indulge in just a little more speculation. Is it possible, we could ask, that UFOs have actually developed some kind of 'invisibility cloak'? If scientists are edging closer to being able to make solid objects invisible, would it not be more than likely that UFOs can do the same?

Take the case of the man who took a picture of the London skyline from a Covent Garden roof, some sixteen floors up, of white elongated 'flying saucer' shapes, complete with humps in the centre. Yet these were not seen at the time and were revealed when the digital picture was examined. We have seen this phenomenon already with the Tantallon Castle 'woman in a ruff' image. The photographer said he was unable to fake it.[77] A triangular and translucent greenish-white UFO was caught by a digital camera in Huddersfield. It was not seen by the photographer at the time, nor by others. The picture was shown in the *Metro* newspaper of 17 August 2009.

But nature occasionally gives glimpses of how this can be done on Earth. Take the case of a drop of dye placed in a viscous fluid such as glycerine which is encased between two glass cylinders, one inside the other. As the outer one is rotated slowly the drop of dye spreads out into the liquid, and soon appears to have totally disappeared with the repeated turning of the cylinder. But if you rotate the cylinder the other way, the drop reconstitutes itself! If we did not know beforehand that the dye drop had been wound into the glycerine, it would seem the particle had miraculously appeared out of nowhere. Similarly, seeing as such a phenomenon exists on Earth, we can postulate that the UFO can do the same; it can 'wind itself' into invisibility, and just as easily unwind itself back into visibility! The drop of dye in the glycerine doubles as a higher-dimensional reality. The atoms that go to make up the UFO, like the dye in the glycerine, are expressions of implicate orders in the universe. Everything is part of the Whole, or the One, and on occasion can be seen to be both a spot or particle by itself or just as easily as part of the Whole.

There are other scientific advances that help us to understand how UFOs can, in an instant, become invisible. 'Permitivity' is the key – the electromagnetic wave can sometimes 'permit' light waves to be seen, or cause them to be deflected away from human vision. Take the shorter colour frequencies often seen in UAPs. These might be part of the normal world, and easily replicated with our own earthly technology.

In fact, scientists could be on the verge of developing an 'invisibility cloak', beloved of sci-fi and fantasy films. They can emulate nanotechnology, where new materials are built up atom by atom. Theoretically, a material can be designed that has the opposite effect – to cancel out light impinging on an object. A 'metamaterial' could mimic the way atoms interact with light. They can be tuned so that the photon skims around the edge of the material of one object placed next to another. The light then moves into a region of the metamaterial that creates a distortion as if the second object were present. The result is that an observer looking at the first object would only see the second object.[78]

It is possible that the US military has developed some kind of invisibility device. William Thomas writes that experimental 'cloaking' techniques can render military jets invisible to ground observers. One woman in New Jersey said she saw 'chemtrail' planes simply 'disappear', although they were in plain sight in clear skies and flying relatively low. A former US military technician spoke of 'Active Camouflage' from patents held by Hughes Aircraft, Raytheon, E-Systems and other military corporations. He explained how images of sky and clouds are sent through fibre-optic cables when 'Active Matrix Liquid Crystal Display' panels were bonded to the opposite side of the aircraft. This projects the illusion – to people on the ground – of looking through the plane.[79] This could explain how UFO investigator Marshall Barnes watched, while driving a car, a fighter jet streak across the highway near Wright Patterson Air Force Base in 1994. The plane started to glow, and then suddenly 'blinked out'. Possibly, by keeping his eye on the road, Barnes missed a vital manoeuvre that took the plane out of his range of vision, or it was hidden by the glare of the sun. He watched to see if it would reappear: of course, it didn't.[80]

Physicists know that it is possible for gaseous substances to take on the appearance of solidity under certain circumstances. They know that electrons sometimes seem like liquids and on other occasions like solids, because they can at times be observed dancing around themselves in well-defined steps, and the character of these steps indicates their nature. In liquids, atoms are randomly distributed, but in solids they are positioned rigidly in a lattice.[81]

Things become solid when the cloud of outer electrons that surround the nucleus prevents hard substances from passing through one another. There is an electric repulsion felt by the electrons in the atoms when one solid object smacks into the other and not because of the lack of available space for the electrons to move through. If something is 'solid' this implies it has mass. But on paper – that is, according to our mathematics – particles do not have mass precisely because they are sometimes waves and not particles. But as photons can *gain* mass, other particles seem to *pretend* to gain mass. The constant search at the CERN laboratories on the French–Swiss border for the famed 'Higgs Boson' particle implies that scientists still do not know what 'mass' is, or how particles have mass, or whether even another particle is needed to give mass to other particles!

In theory, the wall of electrons preventing particles from passing through one another could be tampered with. Scientists seem now to be able to overcome the natural repulsiveness of the electrical and magnetic fields that is inside all atomic structures, at least momentarily.

Laser experiments at Berkeley have succeeded in rearranging, below the electron energy cloud, subatomic particles, which in turn means that in theory, and perhaps in practice, solid objects can be made to pass through or penetrate one another. For example, an X-ray beam was made to fly straight through an aluminium particle before the particle could rearrange itself, making the aluminium technically, if for a split second, transparent to the bombarding X-rays, in the sense that they are not deflected from the hard surface. The experimenters found, to their surprise, that they could easily knock out an electron or two. And if they had continued with the bombardment they would have prevented the missing electrons from replacing one another.

The snag here is the instant re-arrangement of the particles – something we noticed in regard to attempts to create artificial lightning in the sky. To destroy the entire electron-cloud of just one atom, let alone the trillions that would make up the body of any solid observable object, could not possibly be done in laboratory experiments. It is not simply a case of altering the electrical forces between atoms, because this would be a form of alchemy. At the subatomic level the binding energies are a million times stronger than the molecular binding energies. To overcome these energies you would need to heat the atoms up to solar-magnitude heat – or the energy level of a hundred 1-megaton hydrogen bombs.

Even so, when two Russian-born physicists received the Nobel

Prize in Physics in October 2010 their discovery implied that scientists can do amazing things with materials that verge on the paranormal. The two scientists, Andre Geim and Konstanin Novoselov, created a form of carbon known as graphene, in which the atoms can link up like chicken wire, forming a lattice *just a single atom thick*, but which at the same time would be 100 times stronger than steel! Atoms, of course, are invisible to all but the most powerful electron microscopes. The implications are astounding: a lattice spread out across the sky made up of single carbon atoms would be an invisible shield that would destroy any jet plane crashing into it!

Notes

1 The 'Fireball' Riddle

1   See *Nexus* magazine, Apr–May. 1997, p.42.
2   Ibid., p.40.
3   Tom Yulsman, *Earth* magazine, February 1998, p.21.
4   Antony Milne, *Doomsday*, Praeger, 2000, p.107.
5   Robin Sheppard, *The Times*, 7 June 2003, p.18.
6   Eugenie Samuel, *New Scientist*, 20 July 2002, p.88.
7   Ion Hobana and Julien Weverberg, *Ufos From Behind the Iron Curtain*, Bantam Books, 1975, pp.10–30.
8   See *Icarus* journal, DOI: 1016/J. Corus; see also David Chiaga, *New Scientist*, 20 March 2010, p.10.
9   See *UFO Data magazine*, Nov–Dec. 2007, p.16.
10 Eugenie Samuel, *New Scientist*, 18 May 2002, p.15.
11 Hazel Muir, *New Scientist*, 23–30 December 2006, pp.50–1.
12 Amanda Gecta, *New Scientist*, 5 September 2009, p.38.
13 Anna Goslin, *New Scientist*, 7 May 2005, pp.30–4.
14 See *Frontiers* magazine, December 1998, p.38.
15 *Space.com*, entered 21 February 2005; http://www.space.com/science/astronomy/bright.flash.
16 Paul Simons, *The Times*, 8 October 2008, p.67.
17 William Corliss, *Earthquakes, Tides, Unidentified Sounds*, University Press of Colorado, 1983, p.148.
18 Ibid., p.149.
19 Paul Simons, *The Times*, 22 October 2004, p.24.
20 See *UFO Data* magazine, Jul–Aug. 2007, pp.29–31.
21 See *Alien Encounter* magazine, October 1997, p.24.
22 William J. Birnes and Harold Burt, *Unsolved Ufo Mysteries*, Warner Books, 2000, p.8.
23 Ibid.
24 http://www.ufomaps.com/city/colorado+springs, 'Colorado Springs sighting'.
25 See *New Scientist*, 13 May 2000, p.15.
26 http://www.mysticaluniverse.com.ufos-aliens/ufosightings/australia, 'Ballarot, Victoria sightings'.
27 See *UFO Magazine*, Jan–Feb. 2002.
28 See *Geophysical Research Letters*, vol.31, p.41.
29 Matthew Moore, *Daily Telegraph*, 5 February 2009.
30 See *New Scientist*, 31 October 2009, p.7.

31 See *Daily Mail*, 3 April 2010, p.19.
32 See *UFO Data* magazine, Mar–Apr. 2006, p.37.
33 Eugenie Samuel, *New Scientist*, 20 July 2006, p.24.
34 See *Alien Encounter* magazine, October 1997, p.24.
35 See *UFO Reality* magazine, Dec. 1997–Jan. 1998, p.6.
36 See *UFO Data* magazine, Jan–Feb. 2007, p.15.
37 Steve Johnson, *UFO Data* magazine, Mar–Apr. 2007, p.11.
38 http://english.Farsnews.com/newstext.
39 See *Sunday Telegraph*, 28 December 1997, p.11.
40 See *UFO Magazine*, May–Jun. 1999, p.36.
41 *UFO Magazine*, Mar–Apr. 1999, p.34.
42 *UFO Magazine*, February 2004, p.35.
43 *UFO Magazine*, Mar–Apr. 2006, p.37.
44 See *Alien Encounter* magazine, December 1997, p.8.
45 See *UFO Data* magazine, Mar–Apr. 2006, p.37.
46 *UFO Data* magazine, Jan–Feb. 2004, p.45.
47 Austen Atkinson, *Impact Earth*, Virgin Publishing, 1999, p.15.
48 See *The Times*, 17 December 1997, p.15.
49 See *UFO Data* magazine, Mar–Jun. 2007, p.10.
50 See *UFO Magazine*, Mar–Apr. 2004, p.20.
51 See *Focus* magazine, June 1997.
52 Nicholas Redfern, *Covert Agenda*, Simon & Schuster, 1997, p.164.
53 See *Alien Encounter* magazine, December 1997; also *Sunday Times*, 28 December 1997.
54 See *The Guardian*, 29 June 2000.
55 http://www.nzherald.co.nz.
56 http://news.bbc.co.uk/l/hi/Wales/north.
57 See *Daily Mail*, 3 April 2010, p.10.

2 Pummelled from the Skies

1 Hazel Muir, *New Scientist*, 15 October 2005, pp.40–3.
2 See *The Times*, 22 January 2007, p.32.
3 Mark Armstrong, *Astronomy Now* magazine, July 2008, p.48.
4 See *New Scientist*, 17 September 2005, p.13.
5 See *Astronomy* magazine, July 2006, p.24.
6 See *New Scientist*, 31 October 2009, p.7.
7 Robin Lloyd, *Scientific American* magazine, June 2010, p.25.
8 See *Evening Standard*, 4 April 2001, p.20.
9 See *Astronomy* magazine, December 2004, p.41.
10 See *Daily Mail*, 18 August 2004, p.32.
11 See *Daily Mail*, 17 December 1997.
12 See *Daily Mail*, 19 March 2004, p.17.
13 See *Impact*. Spaceguard UK bulletin, no.14.
14 See *Nature* journal, May 2000, vol. 405, p.321.
15 Nigel Hawkes, *The Times*, 18 May 2000.
16 See *New Scientist*, 8 January 2005, p.4; see also *New Scientist*, 25 June 2005, p.36.

17 http://www.space.com; see also http://tinyvol.com/ng3798.
18 See *Impact*. Spaceguard UK bulletin, no.9, November 1999, citing 'BBC online report'.
19 See *Nexus* magazine, Jun–Jul. 2010, citing internet source: 'The Register, co.uk, 28 May 2010'.
20 Brad Steiger, *Mysteries of Time and Space*, Sphere Books, 1977, p.75; see also *Fate* magazine, January 1955; see also B. Rickard and J. Michell, *The Rough Guide to Unexplained Phenomena*, Rough Guides, 2007, p.55.
21 See *Mysteries of the Unexplained*, Readers Digest, 1989, pp.185–6.
22 John Fairley and Simon Welfare, *Arthur C. Clarke's World of Strange Powers*, Book Club Associates, 1985, p.38.
23 Ibid., p.40.
24 Ibid., p.39.
25 Michael Shears, *Daily Mail*, 14 June 2004, p.42.
26 See *UFO Data* magazine, May–Jun. 2007, pp.61–2.
27 Frank Edwards, *Strange World*, Bantam Books, 1973, p.91.
28 Report of the 47th meeting of the BAAS, 1874, p.272.
29 See *American Journal of Science and Arts*, 2:449, November 1857.
30 Philip J. Imbrogno, *Files From the Edge*, Llewellyn Publications (US), 2010, p.43.
31 See *Mysteries of the Unexplained*, pp.184–5.
32 See *The Rough Guide to Unexplained Phenomena*, p.29.
33 Ibid., p.50.
34 See *Edinburgh New Philosophical Journal*, 47:371, 1849.
35 *Mysteries of Time and Space*, p.80.
36 See *Daily Mail*, 23 July 2010, p.3.
37 See *Information Journal*, 1:17–19, spring 1968.
38 See *Los Angeles Examiner*, 5 June 1953.
39 See *The Rough Guide to Unexplained Phenomena*, p.53.
40 Ibid., p.47.
41 Ibid., p.48.
42 *Strange World*, pp.199–200.
43 Aimee Michel, *The Truth About Flying Saucers*, Readers Digest, 1989, p.203.
44 See *Marine Observer*, 49: 17–18, January 1970.
45 *Strange World*, p.192.
46 Brad Steiger, *Mysteries of Time and Space*, p.79.
47 Ibid.
48 *Mysteries of Time and Space*, p.81.
49 See *The Rough Guide to Unexplained Phenomena*, p.48.
50 See *Arthur C. Clarke's Mysterious World*, Thames TV, 1980.
51 *Arthur C. Clarke's World of Strange Powers*, p.40.
52 See *UFO Data* magazine, Mar–Apr. 2006, p.9.
53 See *New Scientist*, 20 September 2008, p.53.
54 Ibid., p.54.
55 Ibid., p.56.
56 Mike Dash, *Borderlands*, Heinemann, 1997, p.249.
57 See *New Scientist*, 20 September 2008, p.56.

58 Steve Johnson, *UFO Data* magazine, May–Jun. 2007, pp.41–4.
59 *Files From the Edge*, p.35.
60 See *Focus* magazine, January 2005, p.43.
61 Ibid.
62 D.A.J. Seargent, *UFOs: A Scientific Enigma*, Sphere Books, 1978 p.41.
63 Ibid.
64 William Thomas, *Chemtrails Confirmed*, Bridger House Publishers, 2004, p.10.
65 Ibid., p.12.
66 Ibid., p.5.
67 Ibid., p.31–32.
68 Ibid., p.38.
69 Ibid., p.47.
70 See *Annals of Philosophy*, 12:93 August 1826.
71 See *Edinburgh Philosophical Journal*, 1:234, October 1819.
72 See *American Journal of Science and Arts*, January 1834.
73 *Mysteries of Time and Space*, p.80.
74 See *Scientific American Journal*, 2:79 28 November 1846.
75 See *Information Journal*, 1:17–19, Spring 1918.
76 *Strange World*, p.214.
77 *Files from the Edge*, p.75.
78 See *Daily Express*, 18 February 2010, p.19.
79 See *UFO Data*, May–Jun. 2006, p.11, and Nov.–Dec. 2006, p.18.
80 http://www.harrowobserver.co.uk, citing Tom Purnell, Harrow Observer, 6 April 2009.

3 The Blackout Mysteries

1   George Foster, *UFO magazine*, October 2003, p.42.
2   See *UFO magazine*, August 2003, p.11.
3   *UFO Magazine*, p.42.
4   David E. Watson, *Nexus* magazine, Dec. 05–Jan. 06, p.11.
5   http://www.nhne.org/news/newsarticles/tabid, citing NAS Report 'How the Sun Could Devastate'.
6   Robin Yapp, *Daily Mail*, 28 October 2003 p.18.
7   See *New Scientist*, 3 February 1996, p.245.
8   Alexei Barrionuevo, *International Herald Tribune*, 12 November 2009, p.18.
9   See *UFO Magazine*, August 2003, citing ECTV universal.heartbreak2010@yahoo.no.
10 Jonathan Leake, *Sunday Times*, 12 January 2003, News section, p.30.
11 Jayne Atherton, *Metro* newspaper, 11 November 2002, p.8.
12 See *Nexus* magazine, Apr–May. 2009, p.29.
13 Frank Edwards, *Strange World*, Bantam Books, 1973 , pp.19–20.
14 Ibid., p.68.
15 John A. Keel, *The Cosmic Question*, Panther Books, 1978, p.53.
16 Gregory M. Kanon, *The Great Ufo Hoax*, Galde Press, 1997, p.47.

17 Michael Hanlon, *Daily Mail*, 20 April 2009, p.15.
18 Steve Johnson, *UFO Data* magazine, Jan–Feb. 2007, p.15.
19 See *The Times*, 3 February 1997.
20 Guy Lyon Playfair and Scott Hill, *The Cycles of Heaven*, Souvenir Press, 1978, p.72.
21 John Carr, *The Times*, 13th July 2004, p.16.

4 Electrical Storm Lights

1 See *Nexus* magazine, Feb–Mar. 2010 , p.47.
2 See *Focus* magazine, July 2000, p.46.
3 Katie Green, *New Scientist*, 25 June 2005, pp.47–49.
4 See *BMJ Journal*, vol. 332, p.1513; see also *New Scientist* 1 July 2006, p.23.
5 See *Nexus* magazine, Feb–Mar. 2010, p.47.
6 See *Nexus* magazine, Oct–Nov. 2005, p.8, citing *Arizona Republic*, 11 August 2005.
7 See *The Times*, online, 4 December 2009.
8 Paul Simons, *The Times*, 22 April 2008, p.64.
9 See *New Scientist*, 7 May 2007, p.30.
10 Hazel Muir, *New Scientist*, 24 June 2006, pp.37–9.
11 Paul Simons, *The Times*, 23 October 2003.
12 See *Daily Mail*, 28 October 2003, p.18.
13 See *Daily Mail*, 13 December 2007 p.31.
14 Hazel Muir, *New Scientist*, 28 June 2003, p.16.
15 Randy Cerveny, *Freaks of the Storm*, Thunder's Mouth Press (US), 2006, p.80.
16 Paul Simons, *The Times,* 20 February 2006, p.63.
17 *Freaks of the Storm*, p.77.
18 Paul Simons, *The Times*, 5 December 2006, p.63.
19 Paul Simons, *The Times*, 26 June 2007 p.65.
20 Paul Simons, *The Times*, 27 April 2004, p.30.
21 Paul Simons, *The Times*, 11 October 2006, p.72.
22 *New Scientist*, 25 June 2005, pp.47–9.
23 Paul Simons, *The Times*, 2 January 2007, p.49.
24 Karen McVeigh, *The Times*, 25 August 2006, p.9.
25 Ibid.
26 See *Daily Mail,* 8 July 2006, p.5.
27 Paul Simons, *The Times*, 11 October 2006, p.72.
28 Simon de Bruxelles, *The Times*, 30 November 2006.
29 See *International Herald Tribune*, 30 August 2008.
30 Don Phillips, *International Herald Tribune*, 1 August 2005, p.1.
31 Paul Simons, *The Times*, 4 August 2005.
32 Paul Simons, *The Times*, 7 October 2008, p.65.
33 *Freaks of the Storm*, p.78.
34 Ibid., p.129.
35 Paul Simons, *The Times*, 11 October 2006, p.72.
36 See *UFO Data*, Mar–Apr. 2006, p.32.

37 See *The Times*, 25 July 2007, p.46.
38 Paul Simons, *The Times*, 20 July 2007, p.73.
39 *Freaks of the Storm*, p.76.
40 See *Daily Express*, 4 August 2009 p.27.
41 Paul Simons, *The Times,* 16 July 2007 p.57.
42 Katie Greene, *New Scientist*, 25 June 2005, pp.47–9.
43 Louis Proud, *Nexus*, Feb–Mar. 2010, p.49.
44 Ibid., p.48.
45 Paul Simons, *The Times*, 20 September 2008, p.71.
46 Louis Proud, *Nexus* magazine, p.45.

5 Our Lit-up Skies

1   Paul Simons, *The Times*, online, 8 June 2009.
2   See *Daily Mail*, 8 June 2006, p.23.
3   David Wilkes, *Daily Mail*, 2 September 2009 p.27.
4   See *UFO Data* magazine, Mar–Apr. 2007, p.54.
5   See *Journal of Geophysical Research*, DOI: 10.1029/2009.
6   See *New Scientist*, 23 September 2006, p.16.
7   Paul Simons, *The Times,* 20 October 2008, p.56.
8   See *Nexus,* Oct–Nov. 2005, p.50.
9   Peter Barker, *Focus*, January 2005, p.46.
10 See *UFO Data* magazine, Jul–Aug. 2006, pp.24–40.
11 Aimé Michel, *Flying Saucers and the Straight-Line Mystery*,
    Criterion Books, 1958.
12 See *UFO Data* magazine, May–Jun. 2006, p.11.
13 See *UFO Data* magazine, Mar–Apr. 2006, p.9.
14 Sarah Knapton, *Daily Telegraph*, 22 March 2009.
15 T. Murray, *Daily Telegraph*, 12 January 2009.
16 David Wilkes, *Daily Mail*, 10 January 2009, p.9.
17 See *UFO Magazine*, October 2003, pp.22–27.
18 See *UFO Data* magazine, Nov–Dec. 2006, p.23.
19 Richard Webber, *Daily Mail Magazine*, 9 May 2009 p.15.
20 Antony Milne files, personal communication.
21 Arthur Shuttlewood, *The Flying Saucerers*, Sphere Books, 1976,
    p.110–11.
22 Ibid., p.112.
23 See *The Marine Observer*, 9:93, May 1932.
24 See *Science: New Series*, 75: 1932, 80–1.
25 See *The Meteorological Magazine*, 71: 134–6, July 1936.
26 See *Natural History* journal, 59: 258–9, June 1950.
27 See *The Marine Observer*, 30: 194, October 1960.
28 D.A.J. Seargent, *Ufos: A Scientific Enigma,* Sphere Books, 1978.
29 *The Flying Saucerers*, p.25.
30 Ibid., p.90.
31 Jenny Randles, *Time Travel*, Blandford, 1994, p.76.
32 Paul Simons, *The Times*, 9 June 2006, p.85.
33 Ibid.

34 See *UFO Data* magazine, Mar–Apr. 2004.
35 Ibid.
36 See *UFO Data* magazine, Mar–Apr. 2006, p.32.
37 See *Nexus* magazine, Oct–Nov. 2005, p.50.
38 See *UFO Data* magazine, Mar–Apr. 2006, p.31.
39 Ibid., p.69.
40 See *UFO Data* magazine, Sep–Oct. 2006, p.28.
41 Ibid., p.30.
42 Ibid.
43 See *UFO Data* magazine, May–Jun. 2007, p.61.
44 See *The Times*, 31 March 2009.
45 See *The Times*, 5 March 2007, p.22.
46 See *Astronomy Now* magazine, May 2007, p.17.

6 The 'Ball of Light' Syndrome

1   Hazel Muir, *New Scientist*, 23–30 December 2006, pp.50–51.
2   See *UFO Data* magazine, Mar–Apr. 2006, p.32.
3   J. Michell and B. Rickard, *The Rough Guide to Unexplained Phenomena*, Rough Guides, 2007, p.150.
4   Paul Simons, *Since Records Began*, Collins, 2008, pp.100–1.
5   See *UFO Data* magazine, May–Jun. 2006, p.11.
6   Paul Simons, *The Times*, 17 February 2009.
7   See *UFO Data* magazine, Jan–Feb. 2007.
8   *New Scientist*, 23–30 December 2006.
9   Karl Hoeber, *Catholic Encyclopaedia*, vol. 8, Robert Appleton Co., 1919.
10 Philip J. Imbrogno, *Files From the Edge*, Llewellyn Publications (US), 2010, p.27.
11 See *Mysteries of the Unexplained*, Readers Digest, 1989, p.100, citing *Complete Book of Charles Fort*.
12 See *Quarterly Journal of the Royal Meteorological Society*, 13:305, October 1887.
13 See *Nature* 12:204 1 July 1880.
14 Brad Steiger (ed.), *Project Blue Book*, Ballantine Books (US), 1976.
15 Paul Simons, *The Times*, online, 18 September 2009.
16 Paul Simons, *The Times*, online, 27 January 2007, citing *International Journal of Meteorology*.
17 See *UFO Data* magazine, Jan–Feb. 2007, p.9.
18 See *Fortean Times* magazine, November 2006, p.24.
19 See *UFO Data* magazine, Jul–Aug. 2006, p.55, citing Dr Peter Sturrock.
20 See *UFO Data* magazine, Nov–Dec. 2007 p.61.
21 'Index', *Beyond Belief*, Orbis Publishing, 1996, p.148.
22 See *Scientific American* magazine, 44:329, 21 May 1881.
23 See *American Journal of Science & Arts*, 2:5, pp.293–4, May 1848.
24 William Corliss, *Earthquakes, Tides, Unidentified Sounds and Related Phenomena*, University Press of Colorado, 1983.

25 Paul Simons, *The Times*, 2 January 2009.
26 *Strange World*, p.173.
27 See 'Index', *Beyond Belief*, p.149.
28 Albert Budden, *UFOs: Psychic Close Encounters*, Blandford, 1995, p.78.
29 See *Weather* journal, 19:228, July 1964.
30 See *Resonance* journal, no. 24, May 1992.
31 *Files from the Edge*, p.32.
32 Tony Taylor, *Fortean Times* magazine, October 2007, p.77.
33 See Paul Devereux, *Earth Lights Revelation: UFOs and Mystery Lightform Phenomena – The Earth's Secret Energy Force*, Blandford Press, 1989.
34 See, Peter Paget, *The Welsh Triangle,* Granada, 1979.
35 See *UFO Data* magazine, Sep–Oct. 2006, p.28.
36 Paul Simons, *The Times,* 15 September 2006, p.80.
37 Paul Simons, *The Times*, 8 April 2008, p.58.
38 Katie Hall and John Pickering, *Beyond Photography*, O-Books, 2006, p.93.
39 Ibid., p.89.
40 See *Beyond* magazine, issue 5, 2007, p.7.
41 See *Since Records Began.*
42 *Strange World*, p.173.
43 Ibid., p.169.
44 Ibid.
45 *Nature* journal, 224:895, 29 November 1969.
46 *Nature* journal, 260: 596–7, 15 April 1976.
47 *UFOs: Psychic Close Encounters*, p.130.
48 Ibid., p.230.
49 See *UFO Data* magazine, May–Jun. 2006, p.15.
50 *Since Records Began*, p.140.
51 Ibid., p.100.
52 Paul Simons, *The Times*, 5 August 2004, p.22.
53 Paul Simons, *The Times*, 23 December 2005, p.67.
54 See *The Times*, 7 September 2007.
55 Paul Simons, *The Times*, online, 8 Jun. 2009.
56 Ibid.
57 See *New Scientist*, 20 March 1993.
58 See *Fortean Times* magazine, March 2007, p.28.
59 Ed Sherwood, *UFO Magazine*, 27 October 2003, p.22.
60 *Encyclopedia of Science and Technology*, McGraw-Hill, 1987.
61 See *Physical Review Letters*, Vol. 96, February 2007.
62 Ibid.
63 Gerry Vassilatos, *Secrets of Cold War Technology,* Adventures Unlimited (US), 2000, p.263.
64 See *Nature* journal, 29 June 1876, pp.193–4.
65 *Mysteries of the Unexplained*, Readers Digest, 1989, p.211.
66 'Index', *Beyond Belief*, p.147.
67 Ibid., p.148.
68 Ibid., p.147.

69   See *Nexus* Jun–Jul. 2009, p.46.
70   See *Fortean Times* magazine, December 2006, p.31, citing *Bulletin of Seismology Society of America*, 63:2177–78, December 1973; see also *Astronomy and Geophysics*, vol. 47/5, p.11–15.
71   See *UFO Data* magazine, Jul–Aug. 2008, pp.32–4.
72   Mike Dash, *Borderlands*, Heinemann, 1997, p.237.
73   Alberto Enriguez, *New Scientist*, 5 July 2003, pp.27–9.
74   'Index', *Beyond Belief*, p.148.
75   Ibid., p.150.
76   *UFOs: Psychic Close Encounters*, pp.140–41.
77   Dan Stober, *San Jose Mercury News*, 8 December 1989
78   Lewis Smith, *The Times*, 7 May 2008, p.28.
79   *Borderlands*, p.238.
80   See *Nature* journal, vol. 403, p.487.
81   *Borderlands*, p.238.
82   See *UFO Data* magazine, Jul–Aug. 2008.
83   Lewis Smith, *The Times*, 7 May 2008, p.27.
84   Chris Hardie, *Nexus* magazine, Aug–Sept. 2001, pp.11–15.
85   Al Baker, *International Herald Tribune*, 24 June 2004, p.3.
86   Chris Hardie, *Nexus* magazine, Aug–Sep. 2001.

7 Folklore Lights and Mirages

1   Paul Simons, *The Times*, 15 March 2007, p.73.
2   Jennifer Westwood and Jacqueline Simpson, *The Penguin Book of Ghosts*, Allen Lane, 2006/08, p.240.
3   Katie Hall and John Pickering, *Beyond Photography*, O-Books, 2006, p.90.
4   Frank Edwards, *Strange World*, Bantam Books, 1969, p.15.
5   Raymond Buckland, *The Weiser Field Guide to Ghosts*, Weiser Books, 2009, pp.155–6.
6   See *UFO Data* magazine, Mar–Apr. 2006, p.42.
7   *The Penguin Book of Ghosts*, p.368.
8   *Beyond Photography*, p.91.
9   Ibid.
10   Philip J. Imbrogno, *Files From the Edge*, Llewellyn Publishers (US), 2010, p.138.
11   Brad Steiger (ed.), *Project Blue Book*, Ballantine Books, 1976, pp.78–100.
12   *Strange World*, p.17.
13   *Files From the Edge*, p.127.
14   Ibid., p.118.
15   *Strange World*, p.16.
16   Ibid.
17   *Files From the Edge*, p.136.
18   *The Weiser Field Guide to Ghosts*, p.143.
19   Ibid., p.154.
20   *Files From the Edge*, p.135.

21 Ibid., p.157.
22 *Strange World*, p.15.
23 *The Weiser Field Guide to Ghosts*, p.154.
24 See *Fortean Times* magazine, January 2007, pp.40–5.
25 See *UFO Magazine*, Nov–Dec. 2003, p.40.
26 Ben Crystal, *New Scientist*, supplement, 10 April 2010, p.vii.
27 See *UFO Magazine*, Nov–Dec. 2003, p.41.
28 Paul Simons, *The Times*, 31 October 2003.
29 See *Mysteries of the Unexplained*, Readers Digest, 1982, p.391, citing *The Complete Charles Fort*.
30 Paul Simons, *The Times*, 17 July 2009.
31 Letter to *The Times*, cited by Paul Simons, 9 April 2009.
32 Paul Simons, *The Times*, 24 October 2008, p.89.
33 Paul Simons, *The Times*, 18 May 2006, p.70.
34 See *UPS Rocky Mountain News*, Denver, 15 May 2006.
35 See *UFO Data* magazine, Jul–Aug. 2008, pp.33–4.
36 Douglas Fox, *New Scientist*, 20 December 2003, p.70, see also *Journal of Optometrists Association of Australia*, also *Independent on Sunday*, 20 April 2003.
37 See *The Times*, online, 7 September 2009.
38 Kris Sherwood, *Nexus*, Apr–May. 2010, p.54.
39 See *Nexus* magazine, Apr–May. 1997.
40 Jay Gourley, *The Great Lakes Triangle*, Fontana Books, 1977, p.15.
41 Boyer, *Great Stories of the Great Lakes*, Dodd, Mead & Co., 1985, pp.179–80, citing US Treasury Dept, 'Wreck Reports', publisher and date unknown.
42 *The Great Lakes Triangle*, p.47.
43 Ibid., p.38.
44 Ibid., p.42.
45 Charles Berlitz, *The Bermuda Triangle*, Panther Books, 1975, p.99.
46 *The Great Lakes Triangle*, p.96.
47 *The Bermuda Triangle*, p.23.
48 Ibid., p.55.
49 See *Mysteries of the Unexplained*, p.237.
50 Richard Winer, *The Devil's Triangle 2*, Bantam Books, 1975, p.64.
51 *The Bermuda Triangle*, p.78.
52 Ibid., pp.76–7.
53 *The Devil's Triangle 2*, pp.66–7.
54 *The Bermuda Triangle*, pp.79–80.
55 *The Devil's Triangle 2*, p.130.
56 See *The Times*, 6 November 2006, p.64.
57 Frank Edwards, *Strange World*, Bantam Books, 1969, p.75.
58 'Index', *Beyond Belief*, Orbis Publishing, 1996, p.100.
59 Ibid., p.100.
60 Ibid., p.102.
61 Ibid.
62 Ibid.
63 Ibid., p.101.
64 *The Devil's Triangle 2*, p.68.

65 *The Great Lakes Triangle*, p.35.
66 Ibid., p.35.
67 See *Daily Mail*, 10 November 2010, p.5.

8 Seeing Things

1 Ben Crystal, *New Scientist*, suppl. vii, 10 April 2010.
2 See *New Scientist*, 7 November 2009, pp.38–9.
3 Timothy H. Goldsmith, *Scientific American*, July 2006, pp.51–7.
4 Fred Attewill, *Metro* newspaper, 18 January 2010, p.25.
5 See *Nature*, vol. 401, p.680.
6 Paul Davies, *Superforce*, Simond & Schuster, 1984, p.48.
7 John P. Briggs and F. David Peat, *Looking Glass Universe*, Fontana, 1985, p.148.
8 Ibid., p.119.
9 Laurence Krauss, *Scientific American*, September 2010, p.19.

9 Camera and Video Evidence

1 Steve Johnson, *UFO Data* magazine, Mar–Apr. 2007, pp.12–18.
2 See *Metro* newspaper, 27 March 2009, p.23, and also *Daily Mail*, 28 March 2009.
3 See *Nexus*, Apr–May. 2009, p.47.
4 See *UFO magazine*, October 2003, pp.40–1.
5 See *UFO Data* magazine, May–Jun. 2006, pp.17–19.
6 Hazel Courtney, *Daily Mail*, 21 July 2007, p.48, reviewing Klaus Heinemann and Miceal Ledwith, *The Orb Project*, Simon & Schuster, 2008.
7 Katie Hall and John Pickering, *Beyond Photography*, O-Books, 2006, p.4.
8 See *Nexus* magazine, Apr–May. 2009, pp.45–9.
9 See *UFO Magazine*, January 2004, pp.50–3.
10 See *Sunday Times* magazine, 31 August 2008, p.50.
11 Philip J. Imbrogno, *Files From the Edge*, Llewellyn Publishers (US), 2010, p.123.
12 See *UFO Data* magazine, Jul–Aug. 2007, p.49.
13 Paul Marks, *New Scientist*, 27 March 2010, p.22.
14 David Pogue, *International Herald Tribune*, 20 August 2009, p.15.

10 The Sky-zappers

1 Gerry Vassilatos, *Secrets of Cold War Technology*, Adventures Unlimited, 2000, p.261.
2 Ibid.
3 Nick Begich and Jeane Manning, *Angels Don't Play This Haarp*, Earthpulse Press, 1995, p.11.

4   See *Nexus* magazine Jun–Jul. 1997, p.37.
5   Gregory M. Kanon, *The Great Ufo Hoax*, Galde Press Inc, 1997, p.43.
6   www.parascope.com.em/1096/testdeth.htm.
7   See *New York Times*, 11 July 1934.
8   See *Nexus* magazine, Dec–Jan. 2001–2, p.45.
9   See International Symposium on Ball Lighting, ISBL97, Tsudawa-Taun, Niigata, Japan, Library of Congress classification QC966.7.B3.S24, 1997.
10 See *Nexus* magazine, Oct–Nov. 2001, p.15.
11 Ibid., p.39.
12 Paul Harris, *Daily Mail*, 24 March 2009, p.30.
13 E.E. Richards, Proceedings, *The Second International Symposium on Non-conventional Energy Technology*, Cadake Publishing, Georgia 30525.
14 See *Daily Mail*, 3 April 2010, p.19.
15 R.C. Hecker, *The APRO Bulletin*, vol. 23, no. 2, November 1973, p.5.
16 Gary Hodder, *Newfoundland Evening Telegram*, 12 April 1978.
17 See *Just Cause* journal, December 1993.
18 See *New Scientist*, 18 April 2009, p.17.
19 *Angels Don't Play This Haarp*, p.41.
20 See US patent 4-959559, 25 September 1990, EP12.
21 See *America's Corporate Families*, 1993, vol. 1, p.552.
22 See *Nexus* magazine, Feb–Mar. 2010, p.24.
23 See *International Herald Tribune*, 11 December 2009, p.3, see also *Daily Mail*, 10 December 2009.
24 See *Nexus* magazine Feb–Mar. 2010, p.25.
25 http://tinyurl.com/. Nexus Feb–Mar. 2010.
26 See *New Scientist*, 18 April 2009, p.17.
27 Harry Mason, *Nexus* magazine, Aug–Sep. 1997, p.39.
28 Tom Yulsman, *Earth* magazine, February 1998, p.21.
29 Harry Mason, *Nexus* magzine, Feb–Mar. 1998, pp.40–3.
30 Ibid., p.44.

11 Man-made Fireballs

1   Antony Milne, *Sky Static*, Praeger, 2002, p.30.
2   See *New Scientist*, 7 April 2007, p.52.
3   See *The Times*, 23 January 1999, p.15.
4   Ibid.
5   See *UFO Data* magazine, July–Aug. 2008, p.32.
6   Radio 5 Live, 2 September 2010.
7   See *UFO Data* magazine, Mar–Apr. 2007, pp.54–6.
8   *Sky Static*, p.27.
9   Jenny Randles, *Something in the Air*, Robert Hale, 1998, p.176.
10 James Davies, *Space Exploration*, Chambers, 1992, p.111.
11 See *UFO Magazine*, Jan–Feb. 2002.
12 Nicholas Redfern, *Cosmic Crashes*, Simon & Schuster, 1999, p.283.

13 Ibid., p.31.
14 Frank Edwards, *Strange World*, Bantam Books, 1969, p.95.
15 Ibid., p.196.
16 Ibid., p.214.
17 See *New Scientist*, 10 July 2010, p.7, citing Union of Concerned Scientists.
18 Jack Challoner, *Space,* Channel 4 Books, 2000, p.45.
19 Proceedings of the Third Conference on Space Debris, European Space Agency, Darmstadt, SP-473, p.262.
20 Paul Marks, *New Scientist*, 31 October 2009.
21 *Space Exploration*, p.191.
22 See *Frontiers* magazine, September 1999, p.38.
23 See *Quest* magazine, October 1998, p.60.
24 Proceedings of the Third Conference on Space Debris, p.207.
25 See *Quest* magazine, October 1998, p.53.
26 See *The Times*, 8 September 2000, p.20.
27 See *Astronomy* magazine, December 2000, p.56.
28 *Cosmic Crashes*, p.292.
29 See *Frontiers* magazine, January 1999, p.101.
30 See *Orbital Decay Quarterly News*, online, February 2000.
31 See *UFO Data* magazine, May–Jun. 2007, pp.45–7.
32 See *UFO Magazine*, Jul–Aug. 1997.
33 See *New Scientist*, 22 January 2000, p.18.
34 See *Astronomy* magazine, December 2000, p.56.
35 See *UFO Magazine*, Jul–Aug. 1997.
36 See *Quest* magazine, October 1998, p.60.
37 *Space*, pp.113–4.
38 See *New Scientist*, 1 September 2001, p.11.
39 See *New Scientist*, 14 November 1998, p.41.
40 *Sky Static*, p.53.
41 Ibid., p.28.
42 Robbie Graham, *Fortean Times* magazine, May 2007, p.56.
43 www.mossnews.com/news/2006/12/01/ufocrash.shtml.
44 See *Orbital Decay Quarterly News*, April 2007, pp.2–3.

12 Mystery Sounds

1 See BBC internet news, 'Have You Heard the Hum?'.
2 Frank Edwards, *Strange World*, Bantam Books, 1969, p.203.
3 http://www.scientificexploration.org/journal/jse, citing Journal of Geosciences, David Deming.
4 http://news.bbc.co.uk/1/hi/england/suffolk/7571870.stn, citing BBC News 'Mystery Surrounds Humming Noise'.
5 http://en.wikipedia.org/wiki/thehum, citing 'The Hum'.
6 See *The Guardian*, 22 July 2004.
7 http://www.scientificexploration.org/journal/jse, op cit.
8 Ibid.
9 Mariane Power, *Daily Mail*, 27 October 2009, p.47.

10   http://www.guardian.co.uk/education/2001,oct/18/medicalscience, citing 'What's That Noise?'.
11   http://www.scientgificexploration.org/journal/jse, citing David Baguley, Journal of Geociences.
12   op cit.
13   See *International Herald Tribune*, 7 January 2009, p.8.
14   Ed Mitchell, *From Headlines to Hard Times*, John Blake, 2009, p.139.
15   See *Quest magazine*, April 1999, p.76.
16   Peter Paget, *The Welsh Triangle*, Granada, 1979.
17   Albert Budden, *Psychic Close Encounters*, Blandford, 1995, pp.114–5.
18   Paul Simons, *The Times*, 27 November 2006, p.54.
19   Laura Burton, *The Guardian*, 18 October 2001.
20   http://news.bbc.co.uk/1/hi/uk/8056284, James Alexander, BBC News.
21   John A. Keel, *The Cosmic Question*, Panther, 1978, p.62.
22   http://www.scientificexploration.org/journal/jse, citing Journal of Geosciences, David Deming.
23   See *UFO Magazine*, August 2003, p.18.
24   http://www.gsl.net/www.taoshum.html, citing 'The Taos Hum'.
25   Ibid.
26   http://www.scientificexploration.org/journal/jse, op cit.
27   Ibid., citing David Deming and Joe Mullins.
28   Ibid.
29   Paul Simons, *The Times*, 22 August 2008.
30   See *Geophysical Research Letters*, DOI: 10/1029/2007.glo33125.
31   Guy Lyon Playfair and Scott Hill, *The Cycles of Heaven*, Souvenir Press, 1978, p.134.
32   See *Nexus* magazine, Apr–May. 2006, p.18.
33   Paul Simons, *The Times*, 26 February 2007, p.59.
34   See William Corliss, *Earthquakes, Tides, Unidentified Sounds and Related Phenomena*, University Press of Colorado, 1983, p.177.
35   Ibid., p.178.
36   Ibid., p.163.
37   Paul Simons, *The Times,* 5 February 2007, p.57.
38   Hugh Hunt, *New Scientist*, 18 April 2009, p.69.
39   Fairley and Welfare, *Arthur C. Clarke's World of Strange Powers*, Book Club Associates, 1985, p.139.
40   Bernard Lagan, *The Times*, 20 November 2006, p.35.
41   John C. Corefield, letter, *Focus Magazine*, June 2008, p.100.
42   See *Fortean Times*, issue 19, p.21, issue 23, pp.29–32, 35–6, issue 24, pp.47–50, issue 93, p.80.
43   *Earthquakes, Tides, Unidentified Sounds and Related Phenomena*, p.141.
44   Albert Budden, *UFOs: Psychic Close Encounters*, p.139, citing Kevin Cunningham, Security Officer, Blackpool Pier, February 1979.
45   See *Science* journal, vol. 324, 2009, p.1026.
46   Ibid.

47   www.theslowdown.notlong.com.
48   See *Nature* journal, vol. 431, 2004, p.662.
49   Ibid., citing Geophysical Research Letters, Nov 2006–Jan 2007, Hadley Leggett.
50   Stephen Leahy, *Wired* magazine, online, 2 April 2005.
51   See *New Scientist*, 5 September 2009, p.38.
52   *Earthquakes, Tides, Unidentified Sounds and Related Phenomena*, p.153.
53   Ibid., p.142.
54   Ibid., p.142.
55   See *Nature*, 2:46, 19 May 1870.
56   *Earthquakes, Tides, Unidentified Sounds and Related Phenomena.*
57   Ibid., p.155.
58   See *Nature*, 2:25–26, 12 May 1870.
59   *Earthquakes, Tides, Unidentified Sounds and Related Phenomena*, p.161.
60   See *Edinburgh Philosophical Journal*, 31:117, Apr–Oct. 1841.

13   Imprints of the Past

1   Peter Haining (ed.), *The Mammoth Book of True Hauntings*, Constable and Robinson, 2008, p.478.
2   Fairley and Welfare, *Arthur C. Clarke's World of Strange Powers*, Book Club Associates, 1985, p.56.
3   Brad Steiger, *Mysteries of Time and Space*, Sphere Books, 1978, p.98.
4   Ibid., pp.100–1.
5   Raymond Buckland, *Weiser Field Guide to Ghosts*, Weiser Books, 2009, p.89.
6   W. Y. Evans-Wentz, *The Fairy-Faith in Celtic Countries*, Oxford University Press, 1911, p.485.
7   *Weiser Field Guide to Ghosts*, p.45.
8   Jennifer Westwood and Jacqueline Simpson, *Penguin Book of Ghosts*, 2006/08, p.111.
9   Bob Rickard and John Michell, *The Rough Guide to the Unexplained*, Rough Guide, 2007, p.210.
10   *The Mammoth Book of True Hauntings*, p.479.
11   Charles Fort, *New Lands*, (Boni and Liveright, 1923, p.111.
12   *Weiser Field Guide to Ghosts*, p.42.
13   *The Fairy-Faith in Celtic Countries.*
14   *The Mammoth Book of True Hauntings*, p.287.
15   *Penguin Book of Ghosts*, p.29.
16   Ibid., pp.368–9.
17   Paul Simons, *Since Records Began*, Collins, 2008, pp.4–6.
18   *Penguin Book of Ghosts*, p.69.
19   Ibid., p.111.
20   Ibid., p.268.
21   *The Rough Guide to Unexplained Phenomena*, p.258.

22 *Notes and Queries*, 1:9:267, March 1864.
23 Lionel and Patricia Fanthorpe, *Mysteries and Secrets of Time*, The Dundurn Group, 2007, p.133.
24 *The Mammoth Book of True Hauntings*, p.273.
25 *World of Strange Powers*, p.287.
26 *The Mammoth Book of True Hauntings*, p.273.
27 Ibid., p.274.
28 Ibid., pp.276–7.
29 Ibid.
30 http://angelsghosts.com/ghost-plane.html, 'Ghost Plane Stories', citing Sheffield Peaks incident.
31 Ibid.
32 *Weiser Field Guide to Ghosts*, p.82.
33 http://news.bbc.co.uk/1/hi/uk/8056284, citing James Alexander, BBC News.
34 *The Mammoth Book of True Hauntings*, p.288.
35 http://paranormal.about.com/6/2005/08/18, 'Ghost Planes', citing Tony Ingle sighting.
36 Ibid.
37 http://paranormal.about.com/6/2005/08/18, 'Phantom Plane Crashes', About.com.Ufo.roundup, vol. 2, no 3.
38 http://paranormal.about.com/library/weekly/aa062899.htm, citing CRIFO Orbit newsletter, vol. 3, no. 1.
39 Ibid., citing Cincinnati Post, 13 August 1976.
40 Ibid., citing CRIFO Orbit newsletter, vol. 2, no 10, citing publication of 1 June 1956.
41 http://angelsghosts.com/ghost-plane.html, 'Ghost Plane Stories'.
42 http://paranormal.about.com/library/weekly/aa062899.htm, Kenny Young.
43 W.G. Wood-Martin, *Traces of the Elder Saints of Ireland*, publisher and date unknown.
44 W. Bottrell, *Traditions and Hearthside Stories of West Cornwall*, publisher unknown, 1870.
45 *Penguin Book of Ghosts*, p.54.
46 http://www.toptenz.net/top-10-ghost-ships.php, citing 'The Flying Dutchman'.
47 Charles Berlitz, *The Dragon's Triangle*, Grafton Books, 1989, p.101.
48 http://www.toptenz.net/top-10-ghost-ships.php, citing 'SS Valencia'.
49 *The Mammoth Book of True Hauntings*, p.19.
50 Ibid., pp.151–160.
51 Frank Edwards, *Strange World*, Bantam Books, 1969, p.73.
52 *Penguin Book of Ghosts*, p.68.
53 http://www.teslasocietyuy.com/mars.
54 Philip J. Imbrogno, *Files From the Edge*, Llewellyn Books, 2010, p.266.
55 *Mysteries of Time and Space*, p.141.
56 Ibid., p.142.
57 *Files From the Edge*, pp.92–3.
58 Arthur Shuttlewood, *The Flying Saucerers*, Sphere Books, 1976, p.50.

59 *The Mammoth Book of True Hauntings*, p.486.
60 Richard Lazarus, *The Case Against Death*, Warner Books, 1993, p.138.
61 Lyall Watson, *The Romeo Error*, Hodder & Stoughton, 1974, p.176.
62 *The Case Against Death*, p.141.
63 K. Raudive, *Breakthrough*, Taplinger, 1971.
64 *The Romeo Error*, p.177.
65 *Files From the Edge*, p.266.
66 *World of Strange Powers*, p.274.
67 Sam Coates, *The Times*, 20 May 2005; see also *Journal of the Society of Psychical Research*, October 2006, pp.211–24.
68 *The Case Against Death*, p.145.
69 Jenny Randles, *Time Travel*, Blandford Books, 1994, p.56.
70 *The Case Against Death*, p.144.
71 Kit Pedler, *Mind Over Matter*, Thames Methuen, 1981.
72 David Greenberg, *Nixon's Shadow*, W.W. Norton Ltd, 2003.
73 *The Mammoth Book of True Hauntings*, p.103.
74 Ibid., p.286.
75 See *Nexus* magazine, Aug–Sep. 2010, p.61, citing STV Scotland, 4 June 2010.
76 *The Mammoth Book of True Hauntings*, p.284.
77 Ibid., p.92.
78 *The Case Against Death*, p.146.
79 Ibid., p.149.
80 *World of Strange Powers,* p.79.

14 Taking UFOs Seriously

1   See *UFO Data* magazine, Nov–Dec. 2007, p.29.
2   Katie Hall and John Pickering, *Beyond Photography*, O-Books, 2006, p.92.
3   See *UFO Data* magazine, Jan–Feb. 2007, p.38.
4   Christopher Hastings, *Sunday Telegraph*, 22 March 2009.
5   See *UFO Data* magazine, Report 1, p.15.
6   Arthur Shuttlewood, *The Flying Saucerers*, Sphere Books, 1976, pp.79–80.
7   Nicholas Redfern, *Cosmic Crashes*, Simon & Schuster 1999, p.261.
8   See *UFO Magazine*, Jul–Aug. 1997.
9   See *UFO Data* magazine, Jul–Aug. 2006, p.36.
10 http://www.bombardierropreto.com, citing Sao Paulo State sightings.
11 Sasjkia Otto, *Daily Telegraph*, 17 August 2009.
12 Frank Edwards, *Flying Saucers – Serious Business*, Bantam Books,1968, p.98.
13 Ibid., p.6.
14 Lawrence Fawcett and Barry Greenwood, *Clear Intent: The Government Cover-Up of the UFO Experience,* Prentice-Hall (US), 1984, pp.9–11.
15 *Flying Saucers – Serious Business*, p.20.

16 Brinsley Le Poer Trench, *Mysterious Visitors*, 1975, Pan Books, p.89.
17 Timothy Good, *Need to Know*, Sidgwick & Jackson, 2006, p.104.
18 See *UFO Data* magazine, issue 4, p.34.
19 See *Everybody's Weekly*, 11 December 1954.
20 See *Astronautics and Aeronautics*, 0:66–70, July 1971.
21 See Ronald D. Story, *Encyclopedia of UFOs*, publisher and date unknown.
22 See *UFO Data* magazine, Jan–Feb. 2006, p.17.
23 L.J. Lorenzen, *The APRO Bulletin*, vol. 24, No. 12, June 1976, p.6.
24 Jorge Martin, *UFO Report*, Sidgwick & Jackson, 1991, pp.103–5.
25 *Need to Know*, p.337.
26 *Ufo Report.*
27 http://1.video-google-com/videoplay?docid, Jack Decaro discussion.
28 See *UFO Data* magazine, May–Jun. 2007, p.10.
29 See *Sunday Express*, 26 September 2010, p.42.
30 Frank Edwards, *Flying Saucers – Here and Now*, Bantam Books, 1968, p.28.
31 Robert Salas and James Klotz, *Faded Giant: The 1967 Missile Incidents*, mail@cufon.org.
32 See *Sunday Express*, 26 September 2010, p.42.
33 Robert L. Salas, *MUFON UFO journal*, no. 345, January 1997, pp.15–17.
34 See *Sunday Express*, 26 September 2010, p.42.
35 See *National Enquirer*, 8 June 1976.
36 Philip J. Imbrogno, *Files From the Edge*, Llewellyn Books (US), 2010, p.99.
37 See *UFO Data* magazine, Jan–Feb. 2008, pp.60–6.
38 See *Phoenix Gazette*, 13 March 1981.
39 Wm J. Birnes and Harold Burt, *Unsolved UFO Mysteries*, Time-Warner, 2000, p.7.
40 Ibid., p.17.
41 See *UFO Magazine*, January 2004, p.54.
42 See *UFO Data* magazine, Jul–Aug. 2007, p.29.
43 http://www.nicap.org/waves.1989/fullrep.html, citing 'Black Triangle sighting'.
44 See *UFO Magazine*, Oct 2003, p.45.
45 Ibid.
46 http://ufos.about.com/ad/thetriangleufos/19/trianglegallery, citing 'Louisiana Triangle Ufo'.
47 Ibid.
48 See *UFO Data* magazine, Jan–Feb. 2008, p.40.

15 The Worldwide Enigma

1 Richard Hall (ed.), *The UFO Evidence*, NICAP (US), 1964.
2 Michel Figuet and Jean Ruchon, *OVNIs, Le Premier Dossier Complet*, Alan Lefeuvre (Fr), 1979.

3   Olmos Ballester, *OVNIs, El Fenomeno Aterrizaje*, Plaza and Janes (Sp), 1978.
4   See *Nexus* magazine, Feb–Mar. 2009, p.4.
5   See Timothy Good, *Need to Know*, Sidgwick & Jackson, 2006.
6   George Knapp, *UFO Magazine* (US), vol. 14, no. 4, April 1999, pp.24–31.
7   Paul Stonehill, *The Soviet Ufo Files*, Quadrillion Publishing, 1998, p.81.
8   Vadim K. Ilyin, *Mufon UFO journal* (US) no. 403, November 2001, pp.8–9.
9   See *Literaturnaya Gazeti*, 7 November 1990.
10 Wm J. Birnes and Harold Burt, *Unsolved Ufo Mysteries*, Time-Warner, 2000, p.57.
11 Ibid., p.60.
12 *Need to Know*, p.357, citing Turkmenistan Criminal Procedures code, Item 5.
13 Wendelle Stevens and Paul Dong, *Ufos Over Modern China*, Ufo Archives, Tuscon, 1983, pp.119–20.
14 Philip Mantle, *UFO Data* magazine, Jan–Feb. 2008, pp.43–4.
15 See *People's Daily* newspaper (China), 27 July 1985, 9 August 1985.
16 See *Agence France-Presse* (China), 5 November 1998.
17 See *Posta* (Turkey), 17 December 2001.
18 See *Nexus* magazine Oct–Nov. 2005, pp.60–61.
19 See *UFO Magazine*, October 2003, p.33, also p.62.
20 See *UFO Data* magazine, Jan–Feb. 2008, p.65.
21 Timothy Good, *Need to Know*, p.382.
22 See *Agence France-Presse*, 1 August 1995.
23 See *UFO Data* magazine, Jul–Aug. 2007, p.30.
24 See *UFO Data* , citing internet evidence.
25 *Need to Know*, p.296.
26 See *UFO Data* magazine, Jan–Feb. 2008, p.63.
27 Nigel Watson, *Beyond Magazine*, May 2007, issue 5, pp.15–21.
28 http://www.mysteriousaustralia.com/rexufo/ufos-bluemountains.shtml, citing 'Ufos Over the Blue Mountains'.
29 See *UFO Data* magazine, Jan–Feb. 2006, p.37.
30 Raul Oscar Chaves, *The Journal of Hispanic Ufology*, 14 October 2005.
31 Steve Johnson, *UFO Data* magazine, Mar–Apr. 2007, p.11.
32 http://www.infobal.com/centenidos 359635.
33 See *UFO Magazine*, September 2003, p.17.
34 Sherry Basker, *Omni* magazine, vol. 8, no. 4, 1986, p.83.
35 See *New Scientist*, 31 March 2007, p.60.
36 See CAA report, Jan–Apr. 1995, Jan. 1996.
37 See *Metro* newspaper, 5 August 2010, p.11.
38 See *UFO Magazine*, October 2003, p.61.
39 Paul Harris, *Observer,* 6 July 2003; see also *Daily Mail* 4 February 2010.
40 See *Daily Mail*, 19 May 2008, p.34.
41 Antony Milne, *Sky Static*, Praeger, 2002, p.56.
42 Paul Harris, *Observer*, 6 July 2003.

43 Frank Edwards, *Strange World*, Bantam Books, 1969, p.92.
44 See *UFO Data* magazine Jul–Aug. 2007, pp.70–112.
45 Michael Evans, *The Times*, 20 January 2005, p.21.
46 See *Honiton and Exmouth Weekly News*, 2 April 1993, p.441, see also *UFO Data* magazine, Nov–Dec. 2007.
47 See *UFO Data* magazine, Sep–Oct. 2006, p.31.
48 See *Daily Express* 21 September 2009, p.5.
49 See *UFO Data* July–Aug. 2008, p.30.
50 See *Sunday Mirror*, 20 February 1983.
51 Luke Salkeld, *Daily Mail*, 23 March 2009.
52 See *UFO Data* July–Aug. 2007, pp.70–112, and Luke Salkeld, *Daily Mail*, 23 March 2009.
53 See *Daily Express*, 18 February 2010, p.19, citing MoD files.
54 See *UFO Data* magazine, Jul–Aug. 2007, pp.29–36.
55 Ibid., p.12.
56 See *Mail on Sunday*, 28 February 2010.
57 See *UFO Data* magazine, July–Aug. 2008, p.31.
58 See *UFO Data* magazine, Mar–Apr. 2008, p.63.

16 Are UFOs Aliens?

1 Michael Talbot, *The Holographic Universe*, HarperCollins, 1996, p.277.
2 Peter Haining (ed.), *The Mammoth Book of True Hauntings*, Constable and Robinson, 2008, p.487.
3 Kenneth Ring, *ReVision magazine*, 11, no. 4, spring 1989, pp.17–24.
4 Emma Rowley, *Daily Express*, 28 September 2009, p.19.
5 Jacques Vallee, *Dimensions: A Casebook of Alien Contact*, Contemporary Books (US) 1988, p.259.
6 See *UFO Magazine*, October 2003, p.61.
7 See *New Scientist*, 29 March 2008, p.44.
8 Sally Guyoncourt, *Daily Mail*, 24 July 2008, p.3.
9 See *Nexus* magazine, Oct–Nov. 2009, p.60.
10 See *UFO Data* magazine, Mar–Apr. 2008, p.48.
11 Frank Edwards, *Strange World*, Bantam Books, 1969, p.24.
12 See *Daily Telegraph* 19 December 2009.
13 See *Daily Express*, 9 April 2010, p.17.
14 Fiona MacRae, *Daily Mail*, 16 February 2009, p.9.
15 Fred Hoyle, *Diseases From Space*, J.M. Dent, 1979.
16 See *The Times*, 21 October 2008, p.28.
17 Paul Davies, *Are We Alone?*, Penguin, 1995, p.70.
18 *The Holographic Universe*, p.281.
19 *New Scientist*, 24 April 2010, pp.26–7.
20 Ibid., 31 October 2009, p.21.
21 Ibid., 26 July 2008, p.17.
22 See *International Herald Tribune*, City suppl., 15 February 2010; see also *Tribune*, 4 February 2010, p.8 citing *Journal of Experimental Biology*.

23 Mrs M.A. Drake, letter to *Daily Mail,* 3 February 2009, p.56.
24 See *Nexus* magazine, Apr–May. 2009, p.47.
25 *Strange World,* p.201.
26 Rickard and Michell, *The Rough Guide to Unexplained Phenomena,* Rough Guide, 2007 p.242.
27 See *UFO Magazine,* October 2003, p.23.
28 Hazel Courtney, *Daily Mail,* 21 July 2007, p.48.
29 *Strange World,* p.43.
30 Leonard H. Stringfield, *Situation Red: The Ufo Siege!,* Doubleday, 1977 pp.135–6.
31 *Strange World,* pp.26–7.
32 Gerald Armstrong, *UFO Magazine,* October 2003, p.35.
33 D.A.J. Sergeant, *Ufos: A Scientific Enigma,* Sphere Books, 1978, p.52.
34 Dr Peter A. McCue, cited in *UFO Data* magazine, Mar–Apr. 2008, pp.40–1.
35 Malcolm Robinson, *UFO Data* magazine, Mar–Apr. 2007, pp.33–4.
36 See *Fortean Times,* January 2007, p.49.
37 Ian Ross Vayro, *Strange and Mysterious Anomalies,* Joshua Books (Australia), 2008, p.111.
38 Jay Gourley, *The Great Lakes Triangle,* Fontana, 1977, p.44.
39 R.C. Hecker, *The APRO Bulletin,* vol. 23, no. 2 November 1976, p.6.
40 J.J. Hurtak, *Nexus* magazine, Oct–Nov. 2009, p.40.
41 Jack Cohen and Ian Stewart, *What Does a Martian Look Like?,* Ebury Press, p.132.
42 See *Nexus* magazine, Aug–Sept. 2009, p.65 book review, *Dark Matter,* New Age Books, 2008.
43 Allen Hynek, *The UFO Experience,* Ballantine books (US), 1974, p.155.
44 Arthur Shuttlewood, *The Flying Saucerers,* Sphere Books, 1976, p.16.
45 See *UFO Data* magazine, Sep–Oct. 2006, p.31.
46 Antony Milne files, personal communication.
47 See *UFO Data* magazine, May–Jun. 2006, pp.22–3.
48 Robert L. Forward, *Future Magic,* Avon Books, 1988, p.108 and 117.
49 Philip J. Imbrogno, *Files From the Edge,* Llewellyn Books (US), 2010, p.15.
50 Malcolm Robinson, *UFO Data* magazine, Mar–Apr. 2007, pp.33–5.
51 Jenny Randles, *Time Travel,* Blandford, 1994, p.78.
52 See *UFO Data* magazine, Nov–Dec. 2006, p.35.
53 *Time Travel,* p.74.
54 Ibid., p.75.
55 Ibid., p.77.
56 Ibid., p.76.
57 Ibid., p.79.
58 Ibid., p.83.
59 See *Nature* journal, vol. 433, 2005, p.498.

60 Jessica Griggs, *New Scientist*, 25 April 2009, p.20.
61 See *UFO Data* magazine, July–Aug. 2006, p.38.
62 See *UFO Data* magazine, Jan–Feb. 2008, p.43.
63 *Ufos: A Scientific Enigma*, p.48.
64 See *UFO Data* magazine, Mar–Apr. 2006, p.32.
65 Ibid., p.32.
66 *The Flying Saucerers*, p.54.
67 See *UFO Magazine*, October 2003, p.45.
68 See *UFO Data* magazine, Nov–Dec. 2006, p.27.
69 Matthew Moore, *Daily Telegraph*, 5 February 2009.
70 See *UFO Data* magazine, May–Jun. 2006, p.23.
71 Matthew Moore, *Daily Telegraph*, 5 February 2009.
72 See *UFO Data* magazine, Jan–Feb. 2006, pp.39–40.
73 Anil Ananthaswarmy, *New Scientist*, 24 October 2009.
74 See *Physical Review Letters*, vol. 102.
75 David Clarke, *UFO Data* magazine, Mar–Apr. 2007, pp.3–9.
76 David Clarke, *UFO Data* magazine, Jul–Aug. 2007, p.12.
77 See *Daily Telegraph*, 18 March 2009.
78 See *UFO Data* magazine, Jan–Feb. 2006, p.25.
79 See *UFO Data* magazine, Jul–Aug. 2006, p.25.
80 Brinsley Le Poer Trench, *The Flying Saucer Story*, Pan Books, 1980, pp.39–41.
81 See *Nature Physics Journal*: DOI: 10.1038/nphys1341. 2009.

Selected Bibliography

Arthur C. Clarke's Mysterious World, Thames TV, 1980

Atkinson, Austen, *Impact Earth*, Virgin Publishing, 1999

Begich, Nick, and Manning, Jeane, *Angels Don't Play This Haarp*, Earthpulse Press, 1995

Berlitz, Charles, *The Bermuda Triangle*, Panther Books, 1975

Budden, Albert, *Ufos: Psychic Close Encounters*, Blandford, 1995

Cerveney, Randy, *Freaks of the Storm*, Thunder's Mouth Press, 2006

Challoner, Jack, *Space*, Channel-4 Books, 2000

Corliss, William, *Earthquakes, Tides, Unidentified Sounds*, University Press of Colorado, 1983

Fairley, John and Welfare, Simon, *Arthur C. Clarke's World of Strange Powers*, Book Club Associates, 1985

Forward, Robert, *Future Magic*, Avon Books, 1988

Good, Timothy, *Need to Know*, Sidgwick & Jackson, 2006

Gourley, Jay, *The Great Lakes Triangle*, Fontana Books, 1977

Gribbin John, *Timewarps*, J.M. Dent, 1979

Hall, Katie and Pickering, John, *Beyond Photography*, O-Books, 2006

Heinemann, Hans and Ledwith, Miceal, *The Orb Project*, Simon & Schuster, 2008

Imbrogno, Philip J., *Files From the Edge*, Llewellyn Publications, 2010

Milne, Antony, *Doomsday: The Science of Catastrophic Events*, Praeger, 2000

Milne, Antony, *Sky Static: The Space Debris Crisis*, Praeger, 2002

Mysteries of the Unexplained, Readers Digest, 1989

Randles, Jenny, *Time Travel*, Blandford, 1994

Redfern, Nicholas, *Covert Agenda*, Simon & Schuster, 1997

Redfern, Nicholas, *Cosmic Crashes*, Simon & Schuster, 1999

Rickard, Bob and Michell, John, *The Rough Guide to Unexplained Phenomena*, Rough Guides, 2007

Sergeant, D.A.J., *UFOs: A Scientific Enigma*, Sphere Books, 1978

Simons, Paul, *Since Records Began*, Collins, 2008

Steiger, Brad, *Mysteries of Time and Space*, Sphere Books, 1977

Talbot, Michael, *The Holographic Universe*, HarperCollins, 1996

Index